Personal
Robotics

Personal Robotics

Real Robots to Construct, Program, and Explore the World

Richard Raucci

A K Peters
Natick, Massachusetts

Editorial, Sales, and Customer Service Office

A K Peters, Ltd.
63 South Avenue
Natick, MA 01760

Library of Congress Cataloging-in-Publication Data

Raucci, Richard.
   Personal robotics : real robots to construct, program, and explore the world / Richard Raucci.
      p.   cm.
   Includes index.
   ISBN 1-56881-089-X
   1. Personal robotics. I. Title.
   TJ211.416.R38 1999
   629.8'92--dc21   98-51745
                                                                 CIP

Printed in the United States of America
03 02 01 00 99   10 9 8 7 6 5 4 3 2 1

To my wife and kids,
prime robot testers, who don't "really"
mind all of the gadgets in the house!

Contents

Acknowledgments

I would like to thank Alice and Klaus Peters and the editorial staff at A K Peters, for being open to the project and for working with me to its completion. I would also like to thank Rick Schneider, for providing electronics and mechanical assembly support, and Jeff Logsdon, Julia Arlinghaus, Jeannine Hansen, and the rest of the Cybernauts for their encouragement and advice. The robot vendors who provided products for review played a large part in the making of this book by understanding the nature of independent reviews. The many creators of Web sites on robotics and related subjects also deserve a mention for providing so much information to a wide audience. Last but not least, I want to thank my family and my parents for their support.

Introduction

T he field of robotics is very interesting to anyone curious about how organisms (including people) interact with the "real world." A true robot kit allows you to get a closer understanding of how human and animal senses work, and how memory (programming) can be used for specific tasks. Some do both, and others only one of these things, but either is sufficient to separate a real robotics kit from a toy.

What Is a Robot, and What Isn't

It's unfortunate that the market for robots is crowded with toy models with no robotic functions, sometimes residing with real robots in the same manufacturer's line. For example, LEGO has two advanced kits that fit the definition of a programmable robot, but they also sell a Technic Giant Robot Set that is only a model of a robot, with no independent functions or programming capabilities. Robotix sells several "robot" kits that have direct control through a wired remote, but the kits only become real robots when you add an additional programmable control unit (bundled with the large-scale Robot Commander set, in some cases).

You can tell if an advertised robot is the real thing by a few simple criteria. Remember, even if it looks like a robot, there's no guarantee that it is. Read the box, and look for words like "programmable" and "sensor action." It's not currently a violation of truth-in-advertising laws to call a toy a robot, so be aware of what you're looking at.

This is not a robot – it's a remote-control (RC) toy with no robot programming capabilities.

This is a real robot – a programmable car model.

Some manufacturers feature midrange robotics kits with incomplete part sets or unnecessarily complicated instructions. We're avoiding the kits that require major soldering and electronic construction for this book, instead concentrating on robotics functions and programming. We'll only be looking at complete kits with full functionality and relatively easy construction.

Other criteria for selection in this book will be price and functionality. Basic industrial robots can start at $20,000, a high price for a personal hobbyist item. However, one of the top industrial robot manufacturers has begun marketing a smaller entry-level robot (the Pioneer series – see Chapter 13) at a reasonable price (starting under $3000), and similar high-function robots are starting to come close to that price range.

Functionality questions are directed at the robot kits, to see if they have a minimum level of interest. For example, a small robot kit like OWI's Hyper Peppy (which runs forward until it hears a sound or touches a wall with its microphone sensor, and then backs up and turns by shifting a front wheel) is of less interest to us than the OWI Navius, which uses an infrared (IR) detection system to navigate by a route you program on a paper disk.

We're also going to disregard any radio-controlled devices that are controlled like a toy car and have no sensor-based functions or programmability. Some of these gadgets do their best to look like robots, but don't be fooled.

Robot Characteristics

A robot can be seen as a machine with three characteristics:

- Input (sensors or stored information)
- Intelligence (interpretation, brains – effected by the presence of a central processing unit)
- Actuators (controlled motion systems or other output devices)

A robot reacts to input by judging its state with its intelligence, and passes commands to its actuators. For example, a robot can judge where a wall is touching it with a sensor; the input is recorded when the sensor is tripped by the obstacle, and passed to the robot's intelligent processing system, which decides that hitting a wall is a bad idea, and moves the robot by passing a command to the actuators.

You can separate the use of stored information from interpretation by means of intelligence in some cases. Consider the analogy of using two different methods to walk across a room. You could clear your mind and use your senses to navigate around obstacles. This is an example of sensor-based navigation. Or you could close your eyes and rely on your memory of where things are in the room to make the same route. This is an example of using a stored program.

Each of the robots featured in this book will have at least one of the following capabilities:

- Sensor-based navigation
- Onboard programming (more than one program, user-input programs, user-changeable programs)

More advanced units will have both features, and you will be able to see how different capabilities can be programmed into your robots.

What Can a Real Robot Do?

A real robot can

- move along a programmed path, and/or
- react with its environment using sensors.

A real robot can also do a combination of the above tasks – for example, a roaming robot could be programmed to wander about a room and change its course if a sensor hits an obstacle. A two-sensor robot could be programmed to turn to the left if the right sensor hits an object, and to the right if the left sensor does.

The robots in this book fit at least one of these categories, and most will fit both.

Robot Controls

Robots can be controlled in a variety of ways. Here are some common interfaces:

Keypad This is usually on the robot itself. It offers a way to program the robot without having to connect it to a computer. Examples of robots with keypads include the Capsela Programmable Command Center, the OWI WAO II, the Robotix Computer Interface, and the TurboZ, all of which will be discussed in this book.

PC Interface This is usually a parallel or serial cable connection to a host computer that you use to download instructions to your robot. You do the programming using control software provided by the vendor. The software ranges from basic text editing of control files to sophisticated Windows applications. Robots with PC interfaces include LEGO MindStorms, FischerTechnik Mobile Robots, the Rug Warrior Pro™, the Robix™ RCS-6, and the Johuco Phoenix.

Some robots, such as the OWI WAO II, include both a keypad and a PC interface.

Other Some kinds of robots feature novel interfaces, including sensor-based programming and a bar code reader. These are interesting since they provide alternative ways to input commands to your robot. The OWI Navius robot uses a light sensor to read a paper disk that you mark to indicate movement. The Capsela Voice Command robot kit uses a sound interpreter chip to distinguish separate commands. The LEGO BarCode Multiset kit can be used to build robots that are programmed using a bar code reader.

None/Alternate Some robots are preprogrammed for sophisticated behaviors. These include BEAM (Biology, Electronics, Aesthetics, and Mechanics) robots like the Solarbotics PhotoPopper Photovore, and the CyBug. These are chiefly of interest because they can explore certain robotic principles. For example, the PhotoPopper is created from available parts (pager motors, simple capacitors and resistors), and powered by sunlight alone. The CyBug comes with a built-in light avoidance program, and can be upgraded to display hunter or prey behavior.

Programming Methods

The general programming method for a robot is to give it an instruction. For example, you might tell the robot to move its motors forward for two seconds. A turn for a two-motor robot would be activated by the instruction "left motor forward, right motor backward," which would turn the robot to the right. In general terms, this is the way most of the robots in this book will be programmed, with increasing complexity as you add sensors and conditional behaviors to your programs.

Another programming method is the Teach mode. This method uses direct control of the robot (like a remote control [RC] car), but stores the steps you take in a program file. The robot then repeats the exact steps each time you run the program. This is a good way to create a sophisticated movement program without having to figure out the steps yourself, and without the necessity of you inputting program code. Two robots that use this method (along with traditional programming methods) are the Robotix Programmable Computer set and the Robix™ RCS-6. This programming method is used widely in industry, and it's a good idea for robot makers to follow it.

Audience

This book is geared to the hobbyist with a keen interest in robotics and technology. Some experience with electrical circuitry and programming can be helpful, but the structure of the book makes it easy to learn as you go along. The starter robot kits act as an excellent tutorial to robotics principles. The midrange robots and robot kits act as a good way to test your skills, and the higher-end robots can hold your interest for a long time.

You'll also learn about how organisms think in the real world, as you try to build behaviors into your robots.

Book Layout

The book is laid out in three sections.

Section One. Contains robotics experimenter toys and kits ($50–$300). Discusses educational-style robot kits from Learning Curve: Robotix, Capsela, LEGO, and TurboZ.

Section Two. Covers hobbyist robots mostly in kit form, with programmability and sensor-based movement ($50–$700). This section covers OWI Elekit robots, a Parallax Basic Stamp kit, the Rug Warrior Pro™, BEAM robots, and the Johuco Phoenix.

Section Three. Discusses commercial-grade robots coming into the general public school market (under $3000 models). This section covers the Pioneer series.

Each chapter focuses on a specific manufacturer or technology. The robots are field-tested, and their features, functionality, and expansion capabilities are discussed. In addition, information on how to program each unit will be given, and ideas for future projects will be laid out.

Robots and Society: Perception vs. Reality

One of the current problems with personal robotics is the vast amount of media attention that robots have gotten over the last hundred years. People have seen and read about robots in a large number of fictional works, while the real work in robotics is generally downplayed. This has led to a false perception of what a robot is.

Robot History

Robots may have gotten the reputation that led directly to the popular misconceptions of the twentieth century due to the development of automatons in the seventeenth and eighteenth centuries. Consider the high level of achievement of the watchmakers who made clockwork robots. These automatons were able to write, play musical instruments, and perform simple actions. They are examples of true robots because they were programmable (via a system of interchangeable cams, for example, a writing automaton could draw different pictures and write separate sentences).

The high level of skill needed to create these automatons led to their introduction into the upper levels of society. Because of their higher

A writing and drawing automaton from
the eighteenth century.

Marie Antoinette's
Chamber Orchestra.

functions, it was almost a given that they would take an appealing (and familiar) human shape. Thus a robot chamber orchestra introduced to the court of Marie Antoinette took the form of a miniature replica of three musicians playing string instruments (violin, cello, and double-bass), not a box with mechanical stalks coming out of it to play them. Other automatons performed as magicians and did dance routines (see the illustrations).

The Orgel Museum in Kyoto, Japan has an excellent collection of robot automatons of these types. Their functions include acting out scenes, performing acrobatics, and playing musical instruments.[1]

One of the highest-functioning types of automatons were the ones created by Jacques Droz. Droz's machines could play musical instru-

An automaton that plays
a mandolin, circa 1890.

An automaton that
performs magic tricks.

A dancing instructor
automaton.

---

[1] See the Web site (http://www.cjn.or.jp/automata/index.html) for more information, including QuickTime clips of these automatons in action (most are over a century old).

The Droz Writer automaton.

The Droz Writer automaton from the rear, showing the interchangeable cam programming system.

ments and draw phrases and artwork. Unlike the standard single-program automatons, Droz incorporated a programming system into his models. Using a system of cams, Droz's Writer could switch between producing a phrase written in flowing script to drawing Cupid and back again.

The Droz Musician.

A view from the back of the Droz Musician, showing the programmable cam system.

The Droz Musician played a small pipe organ, correctly pressing the keys in order to play different melodies (again, programmed through an interchangeable cam system).

But what happens when you make a humanoid robot without being able to exploit its exterior fully? The maker's attention goes to the robot's task, and its exterior is left static. These popular court and public display robots could do many things, but they had immobile faces and bodies of cold painted metal.

Despite the novelty of the automatons of the past, only the idea of a metal man stuck, as the media noted over subsequent years. The coming of the industrial age, with its heavy use of machinery, had a wide cultural impact. The general sense that technology was running away with itself was first felt during this time. Why not transfer those feelings to automata? People felt that it would be unwise to develop a high-functioning mechanical human with no emotions (and no humanoid behaviors, per se), because it would pose a threat to individual workers (depriving them of jobs), and then to general society (as a part of the fear of technology running amok).

This idea was cemented into place by a play called "R.U.R." (Rossum's Universal Robots), written by Karl Capek circa 1920, which

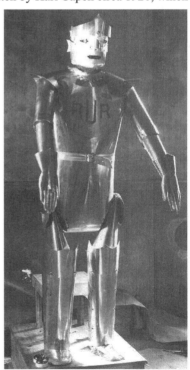

An English robot built for a production of "R.U.R."

expertly summed up these prejudical feelings about automata, and even gave them the slightly derogatory name "robota" (Czech for worker, or drone) which has stuck for all classes of these machines to this day. This wasn't the first time that automatons (prerobots) had gotten a bad rap, but the play certainly helped define the stereotype for the modern age.

The stage was set for "R.U.R." by a number of literary precursors. Most important, although not a story about automata, was Mary Shelley's *Frankenstein* (1834). The complex issues surrounding the creation of a humanoid by scientific means (though stitched together from corpses, Frankenstein's monster is animated by electricity, like robots) are explored here to a great extent. It's unfortunate that the more sensational aspects of the story (i.e., a lumbering hostile monster rampaging across the countryside) are what is remembered.

Nathaniel Hawthorne wrote an excellent story about a craftsman who wants to make non-humanoid "natural" automata in "The Artist of The Beautiful" (a.k.a. "The Mechanical Butterfly"), in 1844. His ethereal butterfly automaton, a complex robot creation of dazzling beauty, is ultimately destroyed, but not before it brings him fulfillment. This is an interesting story in light of the BEAM robotic movement started by Mark Tilden 150 years later.

Herman Melville got back to the robot-as-horror theme by writing "The Bell Tower" in 1856, a story about a humanoid bell-ringer that kills its creator while running along a programmed track.

Jerome K. Jerome's "The Dancing Partner" (1893) is a cautionary tale of a simple programmed machine running amok (in this case, a walking humanoid robot programmed to perform a waltz, with adjustable speed controls and audio output). It's probably the single best argument about not sticking too closely to the human model for making a task-oriented robot.

Ambrose Bierce's "Moxon's Master" (1909), in which an intelligent robot chessmaster kills his maker, is one of the first stories to inveigh a mechanical human with emotion (the robot kills the maker in a fit of fury after being checkmated by the man). It was obviously inspired by the false chess automatons of the 1800s, as outlined in Edgar Allan Poe's "Maezel's Chess-Player" (1836), but Bierce goes a step further by explicitly stating that this was a machine.[2]

---

[2] You can find several sources as well as the electronic text for Mary Shelley's *Frankenstein* on the web at http://www.georgetown.edu/irvinemj/english016/franken/franken.html. The text of Hawthorne's "The Mechanical Butterfly" is accessible at (http://eldred.ne.mediaone.net/nh/aotb.html). Melville's "The Bell Tower" appears at http://www.melville.org/belltowr.htm. And the texts of Jerome's "The Dancing Partner" and Bierce's "Moxon Master" appear at http://www.sff.net/people/doylemacdonald/1_dancing.htm and at http://www.sff.net/people/DoyleMacdonald/1_moxon.htm, respectively.

The robots in Capek's "R.U.R." are suffused with emotion. The entire plot is about a race of intelligent robot slaves working to throw off their shackles. The concept was degraded into the hulking metal monster that appealed to filmmakers in the 1920s–1940s, with a notable exception for Fritz Lang's *Metropolis* (the maria-robot, though a beautiful creation, was still one of the villains—a robot simulacrum of a real woman named Maria, used to keep the workers of the future enslaved).

The maria-robot from Metropolis, 1927.

In 1939, Isaac Asimov started writing stories about robots. Due to the technology of the time, he concerned himself initially with the standard "hulking metal monster" stereotype. A robot babysitter stood nine feet tall and created what Asimov described as the "Frankenstein complex" in human beings. Thus came the Three Laws of Robotics (1942), modeled closely on the robot stereotypes of the nineteenth century, as carried over into the twentieth.

The Three Laws of Robotics are designed to protect humans from robots. They proscribe robot activity to a harsh degree, setting up a rigorous code of behavior that would tax the freedom of any intelligent creature (human being or machine).

They are stated thusly:

1. A robot may not injure a human being, or, through inaction, allow a human being to come to harm.

2. A robot must obey the orders given to it by human beings, except where such orders would conflict with the First Law.

3. A robot must protect its own existence as long as such protection does not conflict with the First and Second Laws.

These are no more than a heavy-handed set of science fiction plot-generating rules, based upon a rather faulty idea of how robots would develop in the latter half of the twentieth century. They spread widely after the Asimov stories were published in the leading science fiction magazine of the day, John Campbell's *Astounding*, and later collected into the anthology *I, Robot* (1960, still in print). Asimov held onto those Laws as plot devices in a number of novels and stories that followed throughout his career.

The Laws show how misconstrued some of the concepts that Capek explored in "R.U.R." were, overshadowed by the concept of automaton-as-menace that preceded him. Rossum, the head of a large manufacturing plant, created his robots to be slave-workers, but they revolted against their conditions and took over the plant. How would Rossum's Universal Robots feel about Asimov's "Three Laws"?

The self-preservation law is the last of the set, and it's subject to the second law, which states that the robot must follow orders. There's nothing about allowing a robot to ignore a human's order to immolate itself to keep the human being warm, for example.

There's also an interesting parallel to H. G. Wells' *The Island of Dr. Moreau* (1896),[3] in which the Laws devised by Dr. Moreau are designed to control a new race of human/animal hybrid creatures. In the book, this is seen as an outrage, forcing the animals against their natures under the threat of a trip to the House of Pain.

If we could increase the intelligence of dogs, would we first lay out a set of servile laws to protect ourselves from them? Would those laws include a prohibition against self-preservation designed to apply to the entire class of dogs?

The real problem with Asimov's Three Laws of Robotics is that they've somehow gained credence (with no experimental evidence) and

---

[3] The text of The Island of Dr. Moreau can be found at http://www.literature.org/Works/H-G-Wells/island/.

are used to plan robots in the real world. This is a surefire way to introduce fantasy into the real-world mechanical, electronic, and software challenges involved in making real robots.

The Robot and the Media

In the past, ideas about technology have affected the progress of its acceptance and implementation, sometimes with disastrous results. What happens when the perception of an invention or product overshadows the reality? Failed expectations make the experience a poor one.

A good example is the airship. After the first successful balloon flights (in the late 1700s), the general prototype for heavier-than-air flight became the airship (basically, a steerable balloon). Everybody knew that the balloon would be the framework behind powered flight. They were wrong, but that didn't stop the media from portraying powered flight as involving an airship for the next hundred years. The idea of the far-ranging airship led to the overutilized airships of the early-to-mid 1900s.

The saturation of the media with the airship led to dangerous preconceptions about their capabilities as far as safety, suitability for different weather conditions, and durability were concerned. The large international programs involving the incredible airships of the 1920s and 1930s led to some spectacular disasters, chiefly centered around mishandling due to poor perceptions. As a result, the entire industry foundered, even though airships were in fact reliable and efficient under proper conditions. They just were incapable of being all-round, all-weather aircraft – definitely *not* the key to "mastery of the air" as the fictional stories and magazine articles led the people in control to believe.

For robotics, the story is somewhat the same. The fictional robot gets in the way of the actual robot, and the actual robot is seen as lacking in capabilities (and, worse, the designs of some robots are cleaved toward an unrealistic mannequin design far beyond their actual capabilities). No, the robots in this book won't walk and talk like a metal human being, but the reality is much more interesting than something like that.

From the 1930s to the 1960s, robots were portrayed as villains more than they were as sympathetic characters. Robots were seen as uncontrollable metal monsters in movie serials, comic books, and magazine articles. A few notable exceptions were the early science fiction character Adam Link and sympathetic robot characters in science fiction stories from Algis Budrys, Isaac Asimov, Ray Bradbury, Clifford Simak, and many others. Notwithstanding these more positive images, the generally accepted idea of robots is that they were dangerous metal figures with stentorian raspy voices.

In 1943, Anthony Boucher took the humanoid robot idea to task in the story "Q. U. R." (Quinby's Usuform Robots, in *hommage* to Karl Capek), in which robots preferred not to have human-type bodies – they wanted to be configured more appropriately to their tasks. It's a prescient view of the robot world from over half a century ago. Currently out of print, the story was featured in one of the first large science fiction anthologies, *Adventures in Time and Space* (Modern Library, 1946; reprinted 1990), as well as the Isaac Asimov anthology *The Golden Years of Science Fiction: Third Series* (edited by Isaac Asimov and Martin H. Greenberg, Bonanza/Crown, 1984).

In the film world, the tin-can robot held sway, most beautifully as Robby the Robot in the film *Forbidden Planet* (1956), designed by Robert Kinoshita, who would later create the familiar Robot (B-9) for the *Lost in Space* television series (1965). Robby operated on a higher plane than the Asimovian robot laws; his main safeguard was an inability to cause harm to an intelligent being (no matter what kind), and being placed in a position to do so set up a dangerous feedback loop that drove him mad (that is, burned out his circuits, in a spectacular display by Disney animators).

Robby the Robot from Forbidden Planet.

The Robot from Lost in Space.

Some interesting alternative robots from the movies include *G.O.G.* (1954), a task-oriented robot on tank treads (who plays tennis and runs a nuclear power plant), and the humanoid robot policeman Gort from *The Day the Earth Stood Still* (1951), who, while still acting as menacing metal human figure, goes beyond this role (he's capable of reason, for example).

The G.O.G. robot from the 1954 film of the same name.

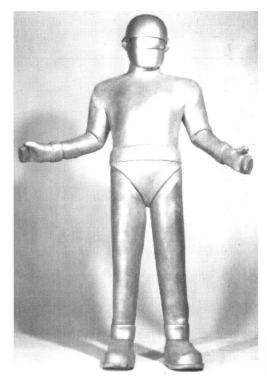

*The robot Gort from
The Day the Earth
Stood Still.*

On television, the robot idea endured as the familiar tin can, or a man in a robot suit, or an actor with a circuit board strapped to his/her back/front from the 1950s until today. Some notable exceptions include a sympathetic version of the 1930s Adam Link robot in an episode of the original *Outer Limits* series ("I, Robot," broadcast 11/14/64), the hu-

*The 1930s science fiction
robot Adam Link, as shown
in the 1964 Outer Limits
episode "I, Robot."*

Mr. R.I.N.G., from Kolchak: the Night Stalker.

manist Robot from *Lost in Space*, and Mr. R.I.N.G., a humanoid robot of high intelligence who escapes from a military project in an episode of *Kolchak: The Night Stalker* (1975). Even though R.I.N.G. is a sympathetic figure, and the story is interesting (the robot is a military project that is set to be deactivated, and he doesn't want to die), he's still obviously a man in a suit – the plot is diminished by the obvious nonrobotic nature of the character.

And who could forget the paranoid robot spaceship in Stanley Kubrick's 1968 film (with Arthur C. Clarke) *2001: A Space Odyssey*? HAL 9000 (the initals are an acroynm for Heuristically programmable ALgorithmic computer) is a high-level multifunction machine intelligence put in charge of all operations of the spaceship Discovery (in effect the ship and its mechanisms become his robot body). He rebels against the human crew when two of them appear to distrust his command of the ship and its highly important space exploration mission. When they actively plot to disconnect him, he kills off four of the crew members and is only prevented from killing off a fifth crew member by a manually

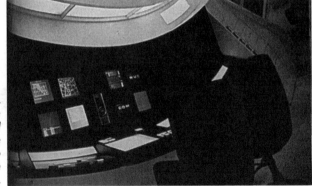

A HAL interface terminal onboard the spaceship Discovery in 2001.

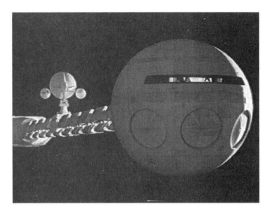

HAL's robot body, the spaceship Discovery.

operated access hatch. The remaining crew member then cuts out his logic circuits in a chilling sequence.

HAL's ability to reprogram himself (essentially, to cause his own systems to report a part failure when there was none to further his own ends – what in humans is called lying) and to operate by his own set of rules asks some interesting questions for future robot developers. In HAL's case, the Asimov Laws "may" have been the proper way to go, as a way to safeguard against such inimical robot mayhem. But as HAL would say, "What about me?"

Nowadays, the public thinks a robot is a walking, talking automaton, along the lines of a television or movie robot (the robot from *Lost in Space*, the droid C-3PO from *Star Wars* [1977]). In the case of the former, seventy-nine episodes of the popular series featured the robot as a major character, complete with a wide range of emotions (even though he had no face). The illusion is persistent – the fact that that robot was essentially a man in a robot suit, with the man doing the voice standing nearby, is lost on a lot of people.

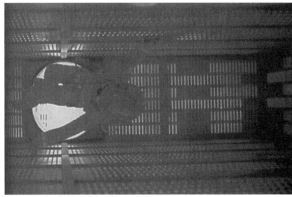

The room containing HAL's CPU (central processing unit) in the film 2001.

The droid robots from Star Wars.

In recent years, more realistic robot designs have developed in films and other media (R2-D2, also from *Star Wars*, is a more feasible robot design, as are the robots (known as drones) from the 1972 film *Silent Running)*.

Humanoid robots are still featured as the norm, however. *Star Trek: The Next Generation's* (1987–94 and several feature films from

A drone robot from Silent Running.

Star Trek: The Next
Generation's
Lt. Commander Data.

1995 on) Lt. Commander Data was a robot far beyond the capabilities of today, but easy to imitate via makeup. The short-lived 1992 series *Earth 2* featured a more realistic humanoid robot named Zero in a small part.

And so it goes, fictional robots engendering a false conception of real robots in the general public.

*Zero, the robot
from* Earth 2.

Real Robots

Real robotics concerns the application of science and mechanics to create a viable robotic mechanism. It is inspired by the fictional robots, but not hampered by the need to live up to an ideal. Since fictional robots (as in films) are created by special effects, they can appear to have increased capabilities that are simply faked. Trying to emulate these capabilities in a real robot will invariably lead to failure.

In the 1960s and 1970s, programmable toys were first introduced. This was paralleled by the initial attempt to produce home robots in the 1970s and 1980s. The early robotic toys were programmed using plastic cams or pegs to control movement. Interestingly, these toys were not sold as robots, but had more robotic capabilities than the false robots being sold in toy stores today.

Mattel's Major Matt Mason Star Seeker with Memory Guidance System, circa 1970, is a prime example. Designed for their line of Man-in-Space toys, the Star Seeker was a small vehicle that was programmed using eleven attached plastic pegs in a circular base. Each peg could be set in one of three positions (left, right, and forward), and a combination

A page from the 1970 Sears Roebuck catalog, showing the programmable Star Seeker.

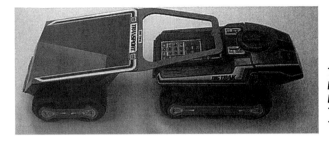

The Milton Bradley BigTrak and Transport Trailer.

of peg placements made up the stored program. Each movement was carried out in order, with up to 22 feet of travel possible. The set also included obstacles around which the user could program the vehicle to navigate.

A more complex robotic toy entry was Milton Bradley's BigTrak, introduced in 1979. This was a programmable robot truck that could move in four directions, with keypad input for programming. The BigTrak also had an optional transport trailer that could be raised and lowered by programming.

Also developing around the time of the BigTrak was the abortive home robot market, circa 1983–85. Apparently a result of the advent of the microcomputer, these robots attempted to be more like fictional robots than live up to their own capabilities. As a result, they ran into poor sales and virtually disappeared from the marketplace by the mid-1980s.

Some of the more notable robots released then include the Hero series from HeathKit, Tomy's OmniBot, the AndroBot, HuBot, and Maxx Steele from Ideal. There were good ideas behind them, but the execution wasn't possible at the time.[4] It's unfortunate that a misconception of the

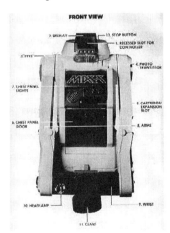

The Maxx Steele robot.

---

[4] See the Web site Robots Wanted (http://robotswanted.com/) for more information on these robots.

A late-model Hero 2000 robot from HeathKit.

Tomy's household robot, OmniBot 2000.

technology needed to make a genuinely useful home robot fifteen years ago would leave that field with virtually no players. It now is apparent that smaller robots following an entirely different path than the humanoid servant model will become the first household robots.

Recent university research developments are bringing robots back from the late 1980s slump, although it would be a shame if robots stayed at the small scuttling research platform stage, or the mechanical insect that walks on six legs, or the demonstration arm. Considering the capabilities of today's technology, the challenges of a complex environment like the average home are within reach.

The robots available today are far from useless; they simply work on a more realistic level. The general ideas are new programming interfaces to make things easier and novel attempts to relate robot behavior to the interaction between sensors. There are interesting ideas about artificial intelligence that you can explore. And making intelligent machines that can act on their own is fun.

You can read about real robots like these in this book. Then get one for yourself!

Section

Educational Robotic Building Kits

Chapter 2

Capsela

C apsela is an electromechanical building set that stresses learning as a part of its appeal. The base unit for Capsela is the capsule, a see-through plastic globe that houses one of the building elements (motors, gears, and special custom parts). The globes attach together with hexagonal connectors. The basic kits have battery-operated motor drives and can be used to build vehicles and simple science experiments. Capsela sets include a wide range of extras – speed reduction gears, crown wheels, internal gear capsules to increase torque, clutches, as well as add-ons like rotary switches, different wheel sizes, treads, gears and chains, special connectors, propellers and fans, floats and impellers, winch and crane units, and small pumps. There are also voice recognition kits that may have special relevance for robotics.

In the mid 1980s, Capsela offered one of the first programmable computers for its building set. The Capsela Computer Program Control CRC 2000 (also available as a unit that could be programmed by IR infrared remote, the ICR 5000) had a battery-backed-up ninety-step single-program computer, which offered nine-direction two-motor control for a robot vehicle, plus ports for a speaker and lights.

The most recent Capsela computer control unit offered is the Programmable Command Center 575, a unique small self-contained programmable robot kit.

The Capsela 575 Programmable Control Center.

Capsela 575 Programmable Command Center

The 575 Programmable Command Center is a robot brain built into a Capsela base unit. It includes four output ports (which connect to robot drive motors), and two dedicated input ports (for a Bump sensor/Light sensor).

It easily assembles into a variety of vehicles, and features a forty-step battery-backed-up programmable memory, an icon-based control keypad, and a full-function liquid-crystal (LCD) status screen. Each program step can be set for up to 7 seconds, giving the unit 280 seconds of maximum programmed movement, but you can also run any program in a continuous loop. This is useful for certain programs where you want your robot to move indefinitely, like a maze-solving routine (with a bump sensor), or a light-tracking program (using the light sensor).

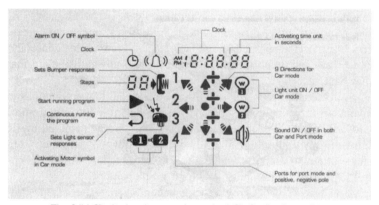

The full LCD display diagram shows the LCD feedback graphics.

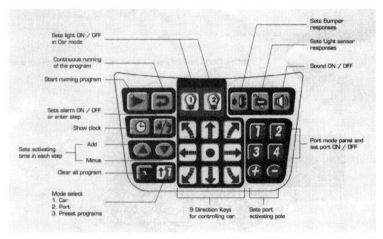

The Membrane Keypad diagram explains the keypad icon-based commands.

Sensor Information

The sensors included with the Capsela 575 kit give this robot kit greater depth as a platform for robotics exploration. Both attach anywhere on the models via standard Capsela hexagonal connectors and can be programmed easily with the keypad. You should, however, make sure the Light Sensor does not project beyond the Bumper Unit if you use them both, since that would interfere with the Bump Sensor.

Bumper Unit

This is a microswitch sensor mounted in a bumper assembly about 4.5 inches long. It acts as a program interrupt and runs a subroutine when triggered (as by striking an obstacle). The Bumper Unit is most useful for programming a wall-following obstacle-avoidance maze-solving program.

Light Sensor

This is a light sensor about 1.5 inches wide in a low-profile package that can be added to any Capsela 575 design. It acts as a program interrupt and runs a subroutine, and can be programmed for either light sensing or dark sensing. You could use this sensor for programming a light-following/light-avoiding program, a daybreak/twilight routine, etc.

Capsela 575 Programming Modes

The Capsela 575 has three programmable modes:

The Car Mode runs according to the command symbols on the keyboard. It also assumes the motors are connected to output ports 3 and

4, and that the vehicle built follows the general plan of the Car models shown in the instructions. Four models are described, but see the building comments – the new shock-absorber wheels tend to get scrunched down under the weight of the 575 unit (with six AA batteries inside it), so we'd suggest Car Mode Combination 2 would work best as assembled, since the motors aren't above the axles.

The Port Mode lets you program and operate the motor capsule functions and light functions separately, or in different positions. This means that all four output ports can be assigned to motors for different functions (fans, vacuums, pumps, winches), as well as for wheels and tracks. This makes the 575 a more versatile unit.

The Preprogrammed Mode offers four preset modes for simple and quick operation. Maybe too simple – the Preprogrammed Mode uses the Car Mode vehicle assumption, and runs four basic programs that show off the unit's range of motion. This is best used to help calibrate the unit running in the Car Mode, or to just show off the robot's controlled-motion routines.

Programming in Car Mode (Mode 1)

To get started, build one of the Car Models, as in the instructions. We recommend #2 for the best results. Make sure the motors are connected to the gears properly, and use the keyboard to input the program.

There are several commands you can program in Car Mode, starting with nine directions for motion, and also including lights and sounds. This is where you'll program the Bumper and Light Sensor responses, as well as the built-in timer/alarm. See the keypad diagram for more information.

After you have built a working robot car model, press the Mode Select key once to enter Car Mode. The LCD display will start with a display of 01, 0 seconds in the top right corner, nine flashing direction symbols, and two flashing motor symbols in the lower left corner. This signals that you can start entering Car Mode commands.

✔ Press any of the Direction keys, Light keys, or the Sound key to program the unit's movement, and the Plus or Minus keys to change the running time for the step (up to 7 seconds, with the default being 1 second).

✔ Press the Alarm/Step key to confirm the step and the program will move to Step 02 (as the display will show). Repeat the process to add steps to your program.

After all the steps are programmed, press the Start key to run the program.

A Basic Car Mode Program:

Square Path

```
01 Forward
02 Right Turn
03 Forward
04 Right Turn
05 Forward
06 Right Turn
07 Forward
08 Right Turn
```

Enter these steps into the robot using the Car Mode display to verify. Hit the Start button to see the robot move through the square path.

You can also use the Continuous key to loop through a program with repeating steps, like the one above. This gives the same result with fewer steps. The Square Path program would then look like this:

```
01 Forward
02 Right Turn
```

Press the Continuous key after you enter the program into memory, then the Start key. You'll see the Loop icon appear on the display. Your robot model should run in a square path until you stop it (by pressing any key).

Further Ideas for Car Mode

You can use the Car Mode to program a maze path for your robot, or a delivery program that includes obstacle avoidance. In the basic programming method, just note the twists and turns of the maze or the delivery path, and program the robot's actions accordingly. We'll be showing how to add sensor-based action to the Car Mode programs in a later section, for even more interesting effects.

Programming in Port Mode (Mode 2)

The Port Mode is used to program the four output ports independently. This means you can create a robot with four-wheel drive, or use the ports for alternative Capsela parts (including fans, pumps, and winches). Note that you may need extra parts for some models. The Capsela 575 manual gives four diagrams of robot vehicles that seem to be more flexible than the Car Mode models. These can be built with the parts in the kit.

Note that Port Mode programming is a bit more confusing than Car Mode. Each output port is assigned a number (1–4) and an action variable using the plus or minus keys (+ or –). Any or all of the ports can be activated in a given step. The + symbol spins the motor forward, and the – symbol moves it backward.

It helps to understand how a robot moves in order to program in Port Mode successfully. The Forward key used in the Car Mode is equivalent to 3 +, 4 + in the Port Mode (with the motors connected to ports 3 and 4). A Right turn in Car Mode means you have to spin the left motor forward (3 +) and the right motor backward (4 -) in Port Mode in the same step.

Use the step-through feature in the Car Mode to observe which motors are running during turns, and you'll get a better idea of how to do the same programming in Port Mode.

Further Ideas for Port Mode Programming

You can create new robots with the full range of Capsela parts. In Port Mode you could build a robot vehicle with four motors, or a vehicle with a raising and lowering winch, as well as a mobile robot that can run a pumping task or a fan under specific conditions.

You can build robots that use the sensors with Port Mode for interesting combinations. For example, you could build a robot that responded to a bright light source (like a candle flame) by activating a small fan (to blow the candle out).

Running the Preprogrammed Mode (Mode 3) Programs

Note that you can't modify the four programs that come already stored in the 575. You'll also have to follow the instructions for building a Car Mode model first before you try running the programs.

To run the programs:

✔ Select the Preprogram Mode (press the program icon three times).

✔ Select the program (1–4) you want to run (use the up and down keys to move between programs).

✔ Hit the Start button.

You can watch the steps as they're executed on the LCD screen. This can help calibration (i.e., if the front left turn goes further than the back left turn, the robot may be overbalanced).

Sensor Programming

How to Set the Bumper Conditions
in the Car and Port Modes

It's easy to program the bump sensor into any program. When the program is running and the bumper makes contact, the bumper preset responses will start (and run to their conclusion). Afterward, the main program will resume.

What this means is that the bumper response will interrupt any program at any given time, run a short subprogram, return the robot to the point where it was interrupted, and then continue the main program.

The Bumper key is used during Car and Port Mode programming to set the bump sensor subprogram. Press it, and you will see the Bumper symbol. Use the same steps to program the bump subprogram as for the main program. To confirm the programming, press the Alarm/Step key and the unit will go back to the previous step in the main program. To switch off the bump sensor responses, press the Bumper key twice, and the symbol on the LCD will disappear.

Steps to follow:

 ✔ Press the Mode key once to set Car Mode, twice for Port Mode.

 ✔ Press the Bumper key once to set the bump sensor subprogram. You'll see the Bumper symbol, and the step indicator will be set to 01.

 ✔ For Car Mode, press any of the Direction keys, Light keys, or Sound keys to program the unit's movement. For Port Mode, use the motor keys and the Plus and Minus keys to pick an action.

 ✔ Use the Plus and Minus keys to change the running time for the step (up to 7 seconds). Then, press the Alarm/Step key to confirm the step. The display will then go back to the main Car Mode/Port Mode program.

 ✔ Finish the program. With the bump sensor connected to the unit, press the Start key to run it.

Observe what happens when the bump sensor encounters an obstacle. It should switch over to the subprogram, and switch back when finished.

Note that the bump sensor subprogram appears to be limited to one step. Plan your programs accordingly for this limitation. For the bump sensor, we were still able to create a good maze-solving obstacle-avoiding program.

Basic Bump Sensor Programs in Car Mode

Obstacle Avoidance Program
Main program:

```
01 Forward (4 seconds)
```

Bump program:

```
01 Back Left (1 second)
```

Use the Continuous setting, and mount the bumper on the front of the robot. The robot will now travel forward until it strikes an obstacle, then reorient itself back and to the left to set up an avoidance path. It will then move forward along the new path, and go around the obstacle.

It's easy to modify the simple program above to create a maze-following robot. By using a wall-following technique, the robot can work its way out of a room or through a maze. The program plan is for the robot to move forward, turn to the wall until it strikes it with the sensor, then make a back turn in the opposite direction (using the bump sensor subprogram), then repeat the process (preferably farther along the wall). Here's what the program looks like:

Wall-Following/Maze-Solving Program
Main Program:

```
01 Forward (3 seconds)
02 Right turn (2 seconds)
```

Bump program:

```
XX Right back turn (1 second)
```

Note that you may have to spend some time creating a maze that will fit your robot, and adjust the times per step in the programs to get the proper action. Capsela units can be finicky. The designs in the 575 manual use two motors to drive the left and right wheels; the alignment isn't generally precise, so one side may pull more than the other, causing the robot to curve when you program it to move in a straight line. You may have to reconfigure the robot to get the best results, and always use fresh nonrechargeable batteries.

You can also use the lights and the sound buzzer in the bump subprogram. For instance, you may want your robot to give off an audible warning when it strikes an obstacle.

Setting the Light Sensor

It's easy to add the light sensor response to any program. When the program is running and the light sensor senses light (or dark), the preset responses will start (and run as long as the sensor is tripped). Afterward, the main program will resume.

This means that the light sensor response will interrupt any program at any given time if used, run a subprogram, then return the robot to the point where it was interrupted and continue the main program.

The Light Sensor key is used in Car and Port Mode programming to set the light sensor subprogram. Press it once and you will see the Light symbol. This indicates that the sensor will react to the absence of light. Press it again to see the Light symbol with light arrows striking it. This indicates that the sensor will react to light.

Use the same steps to program the light sensor subprogram as above. To confirm the programming, press the Alarm/Step key and the unit will go back to the previous step in the main program. To switch off the light sensor responses, press the Light key until the Light symbol on the LCD disappears.

Steps:

✔ Press Mode key once to set Car Mode, twice for Port Mode.

✔ Press the Light key once to set the light sensor subprogram. You'll see the Light symbol, and the step indicator will be set to any main program step. Press it twice for the Light symbol with the light arrows.

✔ Use any direction keys, Light keys (LED light-emitting diodes), or Sound key to program the unit's movement in Car Mode (or the motor keys and the Plus and Minus keys to set the actions in Port Mode). Use the Plus and Minus keys to change the running time for the step (up to 7 seconds).

✔ Then, press the Alarm/Step key to confirm the step and the display will go back to the main Car Mode program.

✔ Finish the program. With the light sensor connected to the unit, press the Start key to run it.

Observe what happens when the light sensor reacts to the absence or presence of light. It should switch over to the subprogram, and switch back when finished.

Note that the light sensor subprogram is limited to one step. Plan your programs accordingly. We were able to create good light-following and alarm programs, even working within the limitations.

Light Sensor Programs

Light Follower

The robot moves in a circle until it detects a light, and then moves forward in the light's direction. If the light is moved to a different spot, the robot goes back to the circling program, picks up the light in the new spot, then follows it.

Here's how to program it:

Main program:

```
01 Forward (2 seconds)
02 Right turn
03 Right turn
04 Right turn
05 Right turn
```

Light sensor program:

(press the Light Sensor key twice)

```
XX Forward (3 seconds)
```

Make sure you position the Light Sensor at the front of the robot, and adjust the times that the steps will take for your particular Capsela robot construction. You want to get the robot to make a complete circle, which can be obtained by turning the robot through one step continuously. Make sure the light levels in your experiment area are low, and use a flashlight with a directed beam.

Light Avoider

Use the same program as above, but modify it to turn the robot away from the light. Although you're only limited to one step for the light sensor subprogram, you can extend it up to 7 seconds. You can also turn on the lights and the sound buzzer in the subprogram. This is more than enough time to spin the robot around in a turn and head it out in the opposite direction.

Dark-Sensing Programs

Light Alert
The robot spins around, sounds an alarm, and flashes its lights when the lights go out.

Main program:

```
01 Stop
```

Dark Sensor subprogram:
```
XX Right turn (7 seconds)
Lights On
Sound On
```

In the continuous mode, the robot should remain still as its main program runs, then run the alarm subprogram when you turn out the lights.

Ideas for further light sensor experiments:

✓ You could use a light sensor program to have the robot look through a series of rooms along a corridor and find the one that is lit up.

✓ Imagine a hallway with four rooms, two on each side. The left and right front rooms are dark. The right back room is also dark, but the left back room is lit. The robot would move forward and sweep around to detect a light source at the junction of each room pair. The first set of rooms would get no response, so the robot would then proceed to the next set (using the same commands). This time, the robot would stop at the left back room and run a light-detection subprogram (for example, moving forward into the lit room, and sounding an alarm to indicate that it found it). The same program should work no matter what room is lit.

Note that this is not suggested as a real-world application. To set this up, you should use a controlled environment (a model of a corridor with rooms in it, for example, sized to your robot).

Multifunctional Programs

Try creating a maze-following program that includes a light sensor sub-program to stop the robot after it leaves a darkened maze. You can also write a program that shows how sensors can help each other. For example, a light-following program combined with obstacle detection would allow an autonomous robot to work its way through a cluttered room to a light source. The light sensor would send the robot toward the light source regardless of the obstacles, and the bump sensor would reorient the robot to get around them.

Capsela Voice Command (VC 3000) and Voice Command 6000

Capsela has also produced a series of voice-controlled robot kits. These robot kits consist of modular control boxes that can interpret spoken words and break them down into executed actions, and motors to drive them about. Voice recognition is an interesting area of robotics because it uses a natural language interface to control the robots you build, without programming. The robot hears the command (to start moving, for example), uses the on-board computer interface to interpret it, and performs the command.

The two voice command sets that come closest to fitting the idea of a real robot are the Capsela Voice Command 3000 (also known as simply Capsela Voice Command) and the Voice Command 6000 set. Note that the Voice Command 6000 may not be readily available at this time.

Capsela Voice Command (VC 3000)

The Capsela Voice Command kit is a basic set that can recognize only a few preprogrammed words. The heart of the unit is the Voice Command Module, which features the voice recognition circuits, a microphone, a battery box, plugs for motor connections, LED lights, a speaker, and standard Capsela connectors. There is also a reset button and a motor mode switch, which you can use to add additional functionality to the set.

The basic words recognized are:

```
Start
Back Up
Turn on the Lights
```

```
Sound Alarm Now
Laser Fire
Raise
Lower
```

The set comes with one motor (which plugs into the A ports on the Voice Command Module), so the basic models you can build are limited to the first five commands. The Raise/Lower commands are used if you add an additional motor to the set (which connects to the B port).

To use the set, build one of the suggested models, and practice saying the commands until you can get the robot to understand you. You'll be limited to making it go back and forth in a straight-line path.

Add a second motor connected to a winch assembly (also available separately), to give the Voice Command set the ability to perform a task. Set the motor mode switch to the B position before you start.

Create a short straight-line delivery path for your robot. At one end of the path, use the winch attached to the vehicle you've built to "Lower" the line /hook, then "Raise" it to pick up a small object. Say "Start" to move the robot to the end point on the path, then "Lower" to deliver the object. Use the "Back Up" command to get the robot back to the starting point (after you remove the object from its path).

This is a fairly limited routine, but it does show how a voice-operated robot works. You could also use the Turn on the Lights and Laser Fire commands along with the motion commands to get the robot to make sounds and flash lights at a particular spot on its path.

You can use the Voice Command Module with other Capsela sets, but the command set will still be limited to what was already described.

Capsela Voice Command 3000 kit.

Capsela Voice Command 6000 kit.

Capsela Voice Command 6000

The VC 6000 is a more advanced version of the Capsela Voice Command 3000. It features a clear plastic dome with the voice recognition circuitry sitting on top of a battery box, which connects to a platform for two drive motors and a set of caster wheels. The control is through a combination of an infrared remote and a series of interpreted voice commands.

The enlarged command set is as follows:

```
Go
Stop
Left Turn
Turn Right
Reverse
Sound
Lights
```

By adding turning commands and two-motor drive to the basic voice command robot, more sophisticated programs can be carried out. For example, you should be able to move the VC 6000 through a maze by directing it with your voice, as well as have it travel more freely about a room.

Special Projects

You might consider the words that a voice-operated robot can recognize as being parts of a program. You could string them together to make repeatable programs. A simple tape recorder with a speaker (or a more sophisticated digital message recorder), mounted onto a Capsela VC vehicle (or placed close to it), should do the trick. Make sure the speaker is facing the microphone, and adjust the volume. You could also try recording your voice as you move the robot about, and playing it back to execute the commands again.

Keep separate tapes for different programs. Think of this as a way to program a robot for specific tasks. In the future, this type of programming could become quite common.

Final Thoughts

Capsela is an innovative electronic building system. While not as complex as some of the robot models that will be described in the following chapters, the 575 has the capabilities for some interesting robotics experiments. The fact that the Capsela 575 and the Voice Command 6000 are both difficult to find, and that the only voice-controlled set in major stores is the limited Voice Command 3000, means that Capsela may not be continuing in the robotics area.

Having said that, the Programmable Command Center is a fine introduction to programmable/sensor-based robotics, if you can find one. Try Young Explorers for the Capsela 575 Programmable Control Center, which retails for around $50, (800) 239-7577.

We'd like to see the Capsela 575 reissued to a wider market. It would be interesting to see a second-generation version, perhaps with a PC interface with downloadable programs.

LEGO
Robotics Kits

L EGO MindStorms is an innovative robotics kit that is centered around a cool robot controller developed by LEGO and the Massachusetts Institute of Technology. The RCX is a programmable CPU set inside a large LEGO brick. It connects to motorized LEGO models to control their robot functions. The kit features 727 LEGO Technic parts to make a wide variety of robots.

The components of the LEGO MindStorms set.

The RCX Programmable Microcontroller

The RCX features three motor connections, three sensor ports, a battery compartment, an LCD status screen, and function buttons (for turning the RCX on and off, running a program, and switching between programs). You can also use an external AC adapter, to save batteries.

You communicate with the RCX through an onboard infrared (IR) receiver/transmitter. The IR transmitter can be used to communicate with other RCX units, for more complex programs. The IR receiver works with another transmitter that attaches to your PC's serial port. You use it to beam programs to the robots (up to five can be stored in memory at one time). The programs are written in the RCX programming language, a graphical user interface (GUI) that allows you to build robot actions and behaviors by connecting LEGO function blocks.

The RCX also features built-in programming to get your models up and running. The five subprograms are

1. Basic motor drive,

2. Move forward and turn,

3. Move forward and stop at light sense,

4. Turn and pause,

5. Move forward and change course (for obstacle detection and avoidance).

The RCX programmable microcontroller.

You use these subprograms with the LEGO parts, motors, and sensors to create the robots you want. The RCX supports these external sensor types:

Touch – a basic bump sensor (On/Off switch)

Light – programmable to detect a range of light levels and wavelengths (i.e., limited color recognition)

Angle – programmable to detect the angle of rotation (in 1/16 increments) of a shaft connected to the sensor

Temperature – programmable to detect a range of temperatures (from –20 to 50 degrees Celsius)

Internal sensor types include:

Counter – programmable to detect a range of actions (for example, the number of times a switch is pressed)

Timer – built-in timer to time actions for conditional programming

RCX – programmed to detect a specific signal from another RCX, then execute a program

In addition to the sensors, the MindStorms RCX line also features

Motors – eight speeds, continuous forward and reverse, plus pulse motion

Lights – eight brightness levels, plus you can set it to blink

Speaker – eight volume levels and two built-in tunes, plus you can program musical notes

The basic set includes one light sensor, two touch sensors and two motors. Additional sensors and other parts can be ordered from LEGO.

Programming the RCX

To program the RCX, use the MindStorms CD application. It gives you a LEGO style programming interface. The five parts are as follows:

Commands (Green)
Sensor Watchers (Blue)
Stack Controllers (Red)
My Commands (Yellow; command macros)
The Vault

The Commands control motor speeds and directions, sound outputs, and related tasks.

The Sensor Watchers monitor the sensors and modify the robot's behavior based on your programming. Up to eight Sensor Watchers can be run in a single program, covering three sensor ports and the RCX IR port.

The Stack Controllers work with the Sensor Watchers to control the robot's behavior. They can be used to set up conditional branches for

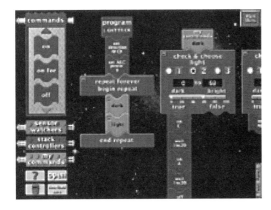

This is a closeup view of an RCX program section.

MindStorms robots. For example, if the light level is low, the robot will return to its base; if it is high, the robot will wander about the room.

The My Commands section is for Command macros that you define. This allows you to set up similar command strings and reuse them without having to put them together each time.

The finished programs are stored in the Vault for reuse.

You make programs by fitting these function blocks together in vertical sets. The general limits for programs are up to eight separate sets, starting with Sensor Watchers, with one Stack Controller and up to twenty commands per set.

The system takes a little getting used to, but the icon-based programming works well, and there are many options for working with sensors, adding conditional programming, and communicating with other RCX units.

Follow the excellent CD-ROM Getting Started Guide for a complete tutorial, including testing the IR interface that you'll use to

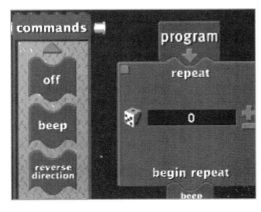

A view of the stacked commands in the RCX programming application.

download programs to your robots, basic building instructions, and programming demonstrations.

Then it's time to get into the Guided Challenges, and see your robot education grow exponentially!

Suggested Basic Models

The basic models you can build with the LEGO MindStorms kit are as follows. They are simple robots that are designed to give you good ideas about how to build the complete robots featured in the Guided Challenges, and also for your own designs. You can find instructions for these models in the Constructopedia manual.

Robo 1 – Slowly moves back and forth using one motor.

Robo 2 – Slowly rotates in one direction using one motor.

Pathfinder 1 – Quickly turns left and right using two motors.

Pathfinder 2 – Slowly turns left and right using two motors.

Acrobot 1 – Quickly turns left and right and does wheelies (pops up on the back wheels) using two motors.

Acrobot 2 – Moves fast and flips upside down using two motors.

The Constructopedia also includes instructions for an advanced robot called TorBot, that you can build with the basic MindStorms set. This one features caterpillar treads for movement and two touch sensors to navigate its environment. The sensor assemblies are built facing downward, and the robot is programmed to use them to find the edges of a table. If the left sensor drops, the robot backs up and turns to the right, and vice versa for the right sensor. In this way, TorBot will patrol the surface of a table, never falling off.

LEGO Dacta RoboLab

As mentioned above, the Pitsco Educational catalog is where you want to go for MindStorms parts. It's also the place where you'll find the RCX classroom alternative, called RoboLab, and the building kits associated with it.

The RoboLab software is a modified version of National Instruments LabVIEW, produced by them in conjunction with LEGO Dacta

*A RoboLab
building set.*

and Tufts University to work with the RCX controller. It's a structured programming interface that's similar to MindStorms, but a bit simpler, and costs $70 separately.

See the Web site for RoboLab at www.natinst.com/robolab/ for more information.

Further Ideas

The LEGO MindStorms set has unlimited potential. See the CD-ROM (compact disc-read-only memory) for robot construction challenges like the Outback Tracker. Use the robot subassemblies found in the Constructopedia to figure out the challenges, and as building blocks for your own creations.

The MindStorms Web site (www.legomindstorms.com) also has ideas you can use for further robotics exploration. Some of the cool robots already created by users are shown in the photo below and on the next couple of pages.

*LEGO MindStorms –
Card Dealer.*

*LEGO MindStorms –
Recycler.*

*LEGO MindStorms –
Copy Machine.*

*LEGO MindStorms –
Mars Rover.*

LEGO MindStorms –
DunkoBot.

LEGO MindStorms –
ATM/Candy Dispenser.

Add-on kits for the basic MindStorms kit are also available. LEGO has produced the Exploration Mars, Extreme Creatures, and RoboSports expansion sets. Each set includes software and a Constructopedia for special robot projects.

The Exploration Mars set allows you build a Mars rover, control it from your PC, and receive sensor readings from the robot. The Extreme Creatures set lets you explore how biological creatures work. It includes parts like light-up fiber-optic strands, pincer and claws attachments, and a wagging tail assembly. The RoboSports kit lets you make robots that play sports. It includes pucks and balls and an extra motor, and instructions on how to build robots that shoot basketballs, play hockey, and run obstacle courses.

*LEGO MindStorms
expansion sets.*

The suggested cost for a MindStorms Expansion Set is $50, and you have to have the basic MindStorms set in order to use them.

Use the RCX to power other LEGO kits. The Technic line of wheeled vehicles is ideally suited for turning into robots. Look for models that have manually steerable wheels. You'll add the motors from the MindStorms set to the steering mechanism and to the drive wheels. You can also add the sensors to help your new mobile robot navigate about.

LEGO's character-based sets could also make interesting robots. For example, you could make a robot Aquazone ship, transform a toy Space Robot into a real mobile robot, or make a robot Star Wars vehicle.

Buy the Technic motor set to get your basic MindStorms set up to three motors. Now you can build more interactive projects. The RoboSports MindStorms Extension Set includes an extra motor as well.

Use two MindStorms RCX robots to show hunter–prey behavior. Put the light on the back of the first robot, and the light sensor on the front of the second. Program the first one to evade obstacles and wander

*An example of a suitable
LEGO Technic kit for use
with the RCX controller.*

One of the robots you can
build with RoboLab.

about, and the second to be attracted to the light. Then watch them move about in tandem.

If you want to get into MindStorms piece-by-piece, see the Pitsco Educational catalog (on the Web at www.pitsco.com). They offer standalone RCX units and transmitters, plus additional school-oriented building kits and an alternative programming interface that runs under Windows and the MacOS.

Final Thoughts

Make no bones about it, LEGO MindStorms is a revolutionary robotics set that will bring real robots to a wide audience. The innovative features of the RCX, combined with the relative ease of use of the programming software (and the great ideas on the CD-ROM) add up to a very positive experience.

The RCX is also very expandable, and other programming interfaces and software have started being developed for it. See the LEGO MindStorms Web Ring for more information (http://members.tripod.com/~ssncommunity/webrings/legoms_index.html).

Lego is also working on a robotics kit for younger users (ages nine and up), which should be available in late 1999. This kit will feature a microcomputer called "The Scout" which will directly control robots

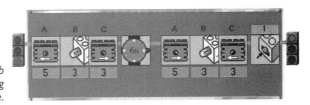

The RoboLab
programming
interface.

constructed from parts in the kit, without a PC. The user will control the robot through a menu system on the onboard computer. For example, pressing the "Go Forward" and "Seek Light" buttons will start a wheeled robot toward the brightest light source in a room.

This system will include the Scout microcomputer, with a built-in light sensor, two touch sensors, two motors, building instructions, and over 400 construction pieces. The suggested retail price is $149. This kit may be a version of the Cybermaster set, sold in the UK, and built around a slightly different version of the RCX.

Contact Information

LEGO MindStorms Set
$219.00

LEGO Systems, Inc.
555 Taylor Road
P.O. Box 1600
Enfield, CT 06083-1600
(860) 749-2291
www.legomindstorms.com

Also available at:

Mondo-tronics Robot Store
4286 Redwood Highway PMB-N #226
San Rafael, CA 94903
(415) 491-4600
info@mondo.com
www.RobotStore.com or www.Mondo.com

LEGO MindStorms should also be available at toy stores like Toys R Us (www.tru.com) and FAO Schwartz (www.faoschwartz.com).

LEGO Dacta Robolab
RCX: $ 115.00
Infrared Transmitter and cable: $25
AC adapter (lets the RCX run off AC power): $25
Robotics Guide and RCX programming software (RoboLab): $70
DACTA Amusement Park Set (293 parts, including two motors, one light sensor, and two touch sensors, as well as building instructions for four RCX-compatible models): $100

Pitsco + LEGO Dacta
P. O. Box 1707
Pittsburg, KS 66762
(800) 362-4308
www.pitsco.com or www.natinst.com/robolab

LEGO Barcode Multiset (Code Pilot)

This robot kit has an innovative programming interface, and makes several interesting (big!) models out of 1200 LEGO Technics pieces. The Code Pilot CPU is a programmable LEGO control brick that fits into most of the example models. It's limited to one motor and one sensor input, and comes with a single motor and two bump sensors.

It features bar code programmable input via a inbuilt scanner (be aware that there's no PC interface for it as yet). The programming steps are printed on a large card that comes with the set, and are also available as small compressed bar codes for the individual modes (in the programming guide). LEGO also provides a software application that lets you create, print, and save Code Pilot bar code steps. Programmable functions include motor power, direction, and duration, sensor actions, and action sounds.

The kit makes an excellent adjunct to the LEGO MindStorms set, as the models that have basic functionality with the Code Pilot would work even better with the RCX.

Suggested Models

Truck

The Loading/Dumping Truck is the largest model in the set. It uses almost all of the parts. The four functions are divided by motor position.

The Truck model sitting on top of the Code Card, used for programming the Code Pilot via bar code scanning.

*The Loading/
Dumping Truck.*

Inserted in the left side, the motor powers the grip arm (which grabs tires in front of the truck, then hoists them over the top and into the back) and the cargo bay (for dumping the cargo and raising it back up). Note that a shift lever is used to separate the two functions, which can't work simultaneously.

In the right position, the motor drives the truck back and forth, with manual steering.

The truck is mainly interesting from a robotics perspective for the way that the arm works. It's an interesting design that uses the single motor to draw the gripper closed, and then raises the arm as a part of the same continuous motion.

Six-Wheel Driver

This big model is also run with the motor in one of two alternate positions. It drives back and forth with the motor in the back position, and raises and lowers its trailer with the motor in the bottom position. It has manual steering and an interesting four-wheel steering assembly. The main drawback is that the model is mostly stationary with a single-motor controller.

*The Six-Wheel
Vehicle.*

The obstacle-avoiding Crash Buggy.

Auto-Drive Crash Buggy

This is the best robot model in the kit, and saves the Bar Code Set / Code Pilot from being less than a true robot kit. This robot car uses a bump sensor to detect obstacles and change course, backing up and turning away when it runs into one, then straightening out its wheels and starting on a new course.

The Crash Buggy can find its way out of a room, and it should be easy to program to find its way out of a maze (by changing the duration and power of the backward movement).

It's a nice one-motor design, that shows how you can streamline a robot for a specific action by paying attention to how you put it together.

The Buggy is also interesting for using as a base model for larger LEGO robots. It's a stable four-wheeled car platform that you can expand.

Robot Walker

This clever two-legged walker has the motor and gear assembly suspended between its legs. It raises a leg, shifts the center of gravity, and

The Robot Walker.

rotates the leg forward and down to make the robot move. Unfortunately, the walker assembly cannot carry the weight of the Code Pilot, so you have to attach it via the motor and sensor cables.

There are also instructions for making a handheld throttle for the robot using the Code Pilot, and for using a sensor switch to make the robot run in reverse. This uses a bar code program to set the speed to the rate of turn of the control, by using the Code Pilot's sensor light to scan the white taco brick (tachometer wheel) attached to it.

Programming Essentials

The Code Pilot is very easy to program. It features direction keys for the motor (front and reverse), a Record button, and the bar code scanner. To enter a program, just press the Rec button until the red light comes on. Then swipe the scanner over the compressed bar code (for a predefined program), or the individual bar code steps from the Code Card, or your Code Pilot software printout. You'll get audible feedback if the program loading was successful.

The program steps are assembled in sequence. In this example, the Crash Buggy is programmed with the steps shown in the following illustration:

The Crash Buggy page from the programming manual, showing the icons that correspond to the bar codes on the Code Card.

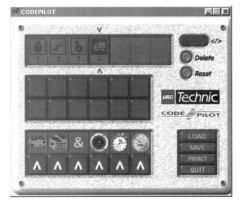

The Code Pilot bar code generating software.

You would use this guide to tell you what to scan from the Code Card; the result is the obstacle-avoiding programmed action. Alternately, you could use the Code Pilot software to enter the steps as listed, and then print a copy to scan. It's easier to correct your mistakes that way, and you can save the file after you create it. Explore www.lego.com for the free software.

The Code Pilot also features a manual programming method (known as Teach Mode). This allows you to store a program as you operate the model from the Code Pilot's keypad. You can then run the program over and over again.

See the Programming Guide for more information, including tutorials.

Further Ideas

Use this kit with the MindStorms RCX controller to create some really impressive robots.

For example, the truck in this set could use three-motor control. There are slots for two motors, but only one can be controlled by the Code Pilot. Put one motor in the left slot, to control the grip arm and the dump function. Put another in the right slot, to control forward and backward movement. Then attach the third motor to the steering control on the top of the truck. You can then program the truck to move about freely, to hunt for cargo, for example, which would then be lifted into the cargo bay when the grip sensor is activated. You'll still need to flip the switch to access the dump function, however, and that might confuse the RCX's programming, so further modifications could be necessary.

The Six-Wheel Driver could also use three motors in the same fashion. That way the drive system, the steering, and the lift arm would all be controllable by a single program.

The Robot Walker model could possibly support the RCX unit as a controller in this fashion: Use a second motor to make an exact copy. Place the RCX unit between the two sets of legs (make sure it's braced properly – you may need to make modifications and order extra parts). You may also be able to use a third motor to steer the modified Robot Walker.

Contact Information

LEGO Barcode Multiset #8479
$164

LEGO Systems, Inc.
555 Taylor Road
P.O. Box 1600
Enfield, CT 06083-1600
(860) 749-2291
www.legomindstorms.com

Learning Curve — Robotix Robot Kits

Learning Curve has developed one of the more interesting educational robotic building sets. The Robotix kits include a Programmable Robotic Computer Set that has instructions for several programmable robots, a full set of Robotix parts to make the models, and a robot CPU brick (the Robotix stand-alone computer), as well as a Robotics Education Kit and Curriculum Guide (that also includes the computer and building parts for robot models), and two full-scale robot building sets (a Robo-Dog robotic dog and a 5-foot-tall Robot Commander mannequin) that can be adapted for programmable control with the same Robotix computer (sold separately).

The heart of the Robotix set is the easy-to-construct modular building system. The parts can be reconfigured to allow you to experiment with your own designs. The sets include detailed instructions for the main models, and alternate models that can be made with the same set.

The motorized Robotix sets use a combination of high-speed and low-speed motors to drive the robots. When used with the Robotix computer, the motors connect to it and are controlled by a stored program. The computer itself is built into a standard Robotix chassis, so you can build a free-ranging robot and easily add it into the other Robotix kits.

The Robotix Computer.

Features of the Robotix Computer

The Robotix Computer is a programmable building brick approximately
8.5 inches long, 3 inches wide, and 2.5 inches deep. It features a top row
of motor connectors (M1–M4) that also have voltage marks (+ and –).
Make sure you have the right marks connected to your cables, or your
motors may turn in reverse.

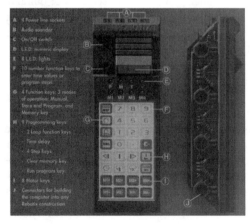

*An overview of the Robotix
Computer's features.*

Directly beneath the motor connectors are the power switch and LED readout screen. The LED screen shows the program step number, the time elapsed during certain functions, and the selectable memory bank

Below the screen are the LED status lights. When lit, P shows that Program Mode is active, M shows Manual Mode, and T shows Trace Mode. The minus symbol shows the motors reversing when lit, turning forward when unlit. When the M1, M2, M3, or M4 status lights are lit, the motors are running or are being programmed.

On the main keypad, there are switches for moving between the computer's three modes of operation. The Program Mode switch sets the programming mode, where you can write and store programs for your robot to follow. The Manual Mode switch turns control to manual, which will turn the motors directly as the motor keys are pressed. This can be used to check the range of motion of your robot before you program it. The Trace Mode switch turns control to a combination of manual and program mode. In Trace Mode, the robot moves directly as a motor key is pressed, but the keys entered are also stored as program steps, and the program can be rerun without pressing the motor keys again. The Trace Mode is probably the easiest way to program your robot, but the Program Mode will give a greater degree of control.

Below the mode keys is the Memory selector. Press this to select the upper, lower, or combined memory bank. The upper and lower banks hold forty steps each, and can be retained as two separate programs. The combined bank holds eighty steps in a single program.

You can use the upper and lower memory banks to assign different program tasks to a single robot. For example, the upper program could follow a maze path, while the lower program could be a pickup and delivery routine.

To the right of the Mode and Memory keys is the number pad, used to input a time value or to go to a specific step. This is also where

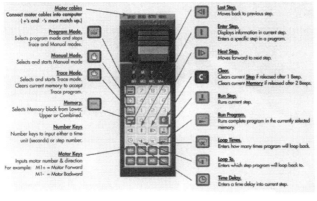

A diagram of the Robotix Computer control panel.

the red Clear key is located. It's used to clear a current step (release it after one beep), or to clear an entire program (release it after two beeps).

Below the number pad are the step command keys. These are used in Program Mode to move between steps, enter information in a step, and run a single step independently. You can run all the steps in a program (in the currently selected memory) with the Prog Run key.

Special programming keys include Loop and Time functions. The Loop Times key enters how many times the program will loop through a series of steps. The Loop To key enters which step the program will loop back to. Basic loop functions allow you to run repetitive steps without having to enter them more than once. See the programming section of this chapter (Using the Computer, below) for more information. The Time Delay key enters a time delay into the current step. Use this to program an inactive period in your program (for example, a period in which the robot is waiting for an event to happen before it starts moving again).

At the bottom of the computer keypad are the Motor keys (forward and reverse, numbered 1 to 4). These are used in all modes to control the robots' motors.

Using the Computer

The Robotix computer manual includes instructions for a basic vehicle that will help you to learn how to program the robots you build. It features two motors: one for rear drive and the other for front-wheel steering. It's a small robotic vehicle that is separate from the computer and connected to it by two motor controlled cables.

The basic vehicle.

Manual Mode

Manual Mode is used to control the robot directly from the keypad. Use the motor keys to initiate movement. The display will show elapsed time since Manual Mode was started, for reference.

Manual Mode is also used to set a model back to a starting position before running a program (so the robot will reach the same endpoint each time). You should also use it to check your robot's range of motion during construction, and to see how well your designs work.

Trace Mode

Trace Mode is a continuation of a time-tested robotics programming method known as Teaching. Robots in industry were run through a program by direct control, and the program was stored in the robot's memory at the same time. This is accomplished by first using the robot's control pad to move the robot about freely (as in Manual Mode).

The robot's actions are recorded as a series of steps in a computer program, that, when run, allows the robot to repeat the same steps. For example, having a robot pick up a part from a shelf and move it to a welder would be programmed by moving the robot's arm using direct control to run it through the actions. After that, the stored program will repeat the learned steps whenever it is run.

This is an easy way to program your robots, since you can see the results as they occur, but you may find it difficult to get a high level of precision without a lot of practice.

Program Mode

Program Mode allows you to enter actions as steps into the robot's memory for precise control. The Robotix computer has an impressive amount of time-per-action capability. You can program up to 999 seconds per step, plus tenths of a second for the first 99 seconds, totaling up to over 16 minutes per step! Note that this time capability will pertain to only certain robots that you build, since the time per step will relate to the range of motion of the robots you build.

To explain this, let's consider the two Robotix kits we'll be working with later in this chapter, the Robo-Dog and the Robot Commander. The Robo-Dog comes with two front-mounted motorized wheels. Running the wheels for the maximum length of a program step could work, since the Robo-Dog would just roll around on its wheels. Using a lengthy step for Computer control of the Robot Commander's arm would be less useful, since the hand open/close time is much shorter, and you don't want to rotate the hand too far (and watch the fingers fly off!). There's a lot of leeway in the programming time, since the theoretical maximum range for a series of forty steps in a standard program could run for about 12 hours.

Use the keypad to enter the program commands. Start by pressing the Program Mode key. The Motor keys are then used with a number value to spin the motors forward and backward for a specific time. This translates into motion for the robots you build – forward/backward/turning drive for a wheeled vehicle, or arm/hand pronation and open/close functions for the Robot Commander. Enter the step, and go on to the next one. Note that you can specify motion for all four motors in one step.

You can check your work with the Step keys, which allow you to edit a specific step and test-run it. This will help you correct programs without having to start all over again.

Use the Loop keys to create subroutines in your programs. This allows you to run repetitive steps without having to write them into the program more than once. An example is to have your robot (for example, the Basic Vehicle described in the computer instruction manual) follow a specific path. A square path, for example, would include a subroutine for a right turn repeated four times. The instruction guide has a clear guide to this process.

You may want to spend some time thinking of how to use the Loop functions with other robots – for example, you could use a Loop subroutine to draw shapes with the Turtle robot in the Projects section of the manual. Going further, think of how to add routines to more complex robots like the Mobile Command Module and the Walking Machine (models you can build with the Computer Set), as well as for the Robo-Dog and the Robot Commander.

The final programming function is the Time Delay key. This allows you to enter a delay step into a program (for example, to stop a robot for a specific length of time, then restart). You can also use this to program a delay into a computer to help you start multiple computers at around the same time (for example, you can program a one-second delay into the first computer to give you time to start the second computer).

Projects

The Computer Set kit has a series of supplemental projects listed in the manual that you can build from the available parts. We'll take a look at the ones that have more interesting robotics features.

The XY Plotter drawing machine.

XY Plotter

The Plotter model uses two motors and a rack-and-pinion gear system to draw and write simple shapes. This type of robotic function goes back to the earliest days of mechanical automatons, clockwork figures that could draw and write.

This drawing machine uses one motor to move along the *x*-axis and another to move along the *y*-axis. The pen is attached to the lower motor rack and draws by being dragged along the paper. See the examples in the manual for some easy shapes to start with, and experiment with your own designs. Use Trace Mode to see what you're drawing and program it into the computer at the same time. For finer drawing control, try motion settings in the Program Mode, which may be set to tenth-of-a-second increments.

For further exploration, you can work on an advanced project that could include adding the *XY* Plotter assembly to a humanoid robot (such as the Robot Commander or a figure that you create). We'd suggest replacing the two-motor grasping arm with a two-motor drawing arm. The drawing tasks you could plan and program for it would add to the appeal of a human-like robot.

Turtle

The robot Turtle is one of the basic small robot shapes that have developed since the 1940s. This simple project uses two motors to drive left and right wheels, and puts a pen in the middle to allow the robot to draw as it moves.

Use the Turtle model to get an idea of how to work with a wheeled robot base. It can help you to see how motor placement affects turning ability.

Further projects could include creating a cargo carrier for the top of the unit. Use the other parts in the Computer Set to make the basic Turtle into a delivery vehicle.

The
Robotix
Turtle.

Robot Arm

The Robot Arm uses a special claw assembly with a motor to open and close it, and a shoulder motor to raise and lower the arm. Use this model to get a sense of how a robot arm works. Check out how the shoulder motor levers the arm forward and backward to position it. See how much control you need to use the claw properly with different small objects. The manual suggests using Trace Mode to move the arm about and enter the program into the computer, but take the time to use Program Mode as well. The finer control will help you to understand how to move a robot arm with a good degree of control.

The basic robot arm is a good addition to a wheeled platform, because it adds functionality (that is, the robot can now go to a spot, pick up an object, and bring it back). Since the computer can control four motors, you can run the arm from two and have another two to drive and steer.

See if you can build a combination vehicle with motors from another Robotix kit, or order additional motors and wheels from the spare parts catalog (you may also need extra cables to connect to the computer).

The Robot
Arm.

The Walking Machine.

Walking Machine

One of the biggest problems robot designers have to face is the complexity of walking as a way of moving about. For humans, animals, and insects, this type of controlled falling comes naturally. For robots, the idea proves a bit harder; wheeled movement is a lot easier to reproduce. But the advantages of walking (the ability to go up steps, for example, and to navigate rough terrain) makes exploring it as an option for your robots worthwhile.

The Computer Set Walking Machine robot uses one motor to shift the legs forward, and another to move a counterbalancing weight. This keeps the robot balanced when one leg is extended and allows it to have enough clearance to slide forward. It's a simple design, and it works well.

Note that the leg/foot assembly won't clear the ground entirely. Another design you could work on would be a walker that could pick up and put down its feet. You might also want to try to make a larger walking robot that could carry the Computer unit on its back, for a self-contained design.

Mobile Command Module

The Mobile Command Module (MCM) uses one motor to drive and one to steer. It's a fully self-contained robot vehicle, built using the computer unit itself as a base. You can program it to move in any direction. The large back roller gives smooth movement, and the steerable front wheel set makes the robot easy to turn.

Use this robot model to get an understanding of how a robot vehicle works. Experiment with path navigation (such as following a maze)

The Mobile Command Module, with the computer built in.

by writing programs for forward, backward, and turning movements. Use the loop functions if you want the robot to repeat motions.

Note that the model you can build with the basic Computer Set has a gripper claw mounted in front, but it's not connected to any motors. You can add two more motors to power the gripper claw (up/down, open/close). Connect the additional motors to the computer, and you're ready to add the claw grip functions to your programs.

You can also construct a payload compartment to allow a vehicle based on the MCM to make deliveries. Use additional Robotix parts to extend the back plate to form an open box. Other designs could include an arm that could pick up and carry objects to this box, then drop them in.

Robo-Dog

Introduction

The Robotix Robo-Dog building set will let you build a full-size robot dog. The main control system included is a wireless remote controller that works with a receiver battery box built into the dog. Four motors power the dog—two for the front wheels, one for the neck, and one for the jaw. Robo-Dog can move forward and backward, make wide turns, open and close its mouth, and turn its head.

Using Robo-Dog with the Computer

Replace the receiver battery box on Robo-Dog with the computer. This is fairly straightforward, since the computer is the same width of the receiver and has connectors in the same spots. The computer will stick out a bit farther than the battery box, and the weight difference isn't that much.

Use the computer to program behaviors for Robo-Dog. For example, Robo-Dog can go from room to room, wagging its head and open-

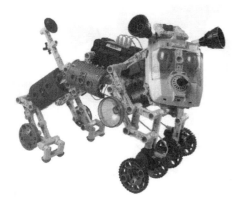

The Robo-Dog.

ing its mouth. Backing the robot up can be difficult, though, because the motors have to carry a lot of weight. You can get the robot to make a large sweeping turn, but you'll need a lot of floor space (like a gym floor).

Experiment with placing the motors, axles, and wheels on the rear legs. You may have to change the overall design a bit. This may help to propel Robo-Dog along. The idea to remember here is that Robo-Dog is designed to look like a full-size robot dog. You could create a very efficient design, but lose the look of the dog. On the other hand, a hybrid robot-dog combination could be very interesting (e.g., you could build a dog with a working hand, or claws).

This is a good way to learn about the concerns of full-scale robot design. The modular Robotix design lets you experiment with the physical construction of your robot for as long as you like.

Other Designs in the Robo-Dog Kit

The Robo-Dog kit also contains instructions for supplemental models you can build, including a Mars Surveyor planetary explorer and a Wreck Truck vehicle. Some of these are especially interesting for use with the Robotix computer. Like Robo-Dog, these models work well with the computer because each uses four motors, the same number that can be controlled by it.

Mars Surveyor

This robot vehicle kit has two-wheel drive and an articulated grip arm. You can raise and lower the arm, and open and close the grip claw to pick up small objects. The precise motor control for the wheels (one motor for each wheel) means that you can steer the robot to a specific location, use the claw, and then move the object you pick up to another location.

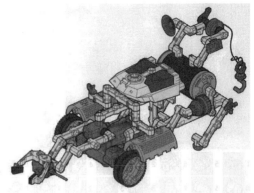

*The Mars Surveyor
basic model.*

To attach the computer, replace the receiver battery box in the building instructions with the computer module. You'll have to make some modifications to the original design to make this work properly. For example, the original design with the receiver unit has a top connector, while the computer does not. To get around this, we used a 50-mm arm piece with a 25-mm inline connector attached to the side of each motor. The 50-mm arms then attach to the sides of the computer. Use a T-shaped angle connector pointing upward between the 50-mm inline connectors to finish the subassembly. The arm shoulder motor will fit into this connector.

Attach the back of the vehicle to the other end of the computer, and remove the lengthwise crossbar assembly (which otherwise would block the keypad). You should now be ready to start programming the robot.

Try a retrieval routine, where the robot will drive to an object, pick it up, and bring it back to you. Then try to program a delivery routine, where the robot would drive to an object, pick it up, and take it to another room (while maneuvering around obstacles). Mars Explorer would also do well as a maze-following robot, running along a preprogrammed path.

Wreck Truck and Probe Launcher

Both of these models may be especially tricky to fit with the computer. You could use the computer alongside them, but that wouldn't be as interesting.

For the Wreck Truck, consider removing the elements that are the most interesting (a motorized boom arm and a separately controlled winch, plus a motorized steering unit) and reassembling them around the com-

The basic Wreck Truck model.

puter unit. You may have to do a fair amount of redesign. Make sure the elements of the robot are balanced, and that the weight isn't all in one spot.

The reason to go to the trouble of adding the computer to the Wreck Truck is that it has alternate capabilities of which you should be aware. This includes a boom arm and a winch with a tow hook. A robotic delivery vehicle may make sense with this arrangement, since the hook can lower to an object with a handle and pick it up without having to grasp it. You can then program it to drive off to its destination.

The Probe Launcher is primarily useful for studying the steering mechanism, which turns the front half of the vehicle (containing the drive motor). The main model has an attached motorized rocket you can raise, lower, and turn, but this feature isn't that interesting for robotic programming.

You may want to disassemble this vehicle as well, and rebuild the interesting parts around the computer. Experiment with this model's steering capabilities, and see how they stack up to other robots you can build.

The Probe Launcher model.

*The Robot
Commander.*

Robot Commander

The Robotix Robot Commander is a unique full-scale humanoid robot that harkens back to the earliest days of robot automatons, but also works well as an up-to-date programmable educational robot (with the addition of the Robotix Computer).

The base set comes with over 400 Robotix pieces, and lets you construct a robot over five feet tall. Control is via a wired remote, which connects to the five motors used in the robot. Two motors are in the right shoulder and hand, and three are in the head (one each for the neck, jaw, and visor/brow). With these motors, your Robot Commander can extend its arm, grasp objects in a five-fingered hand, turn its head, open and close its mouth, and raise and lower its brow. This good range of actions lets you explore the automaton side of robotics.

Automatons are robots that are designed not only to perform tasks, but also to look and act like humans (and animals). These types of robots are usually found in entertainment (as, for example, the programmed robots in Disneyland arcades), but they may also make it into the home, since they could adapt easily to the home environment. For example, a small wheeled robot might not be tall enough to reach a table with a grip claw, but a robot like the Robot Commander could, since its mechanical arm assembly is three feet off the ground.

You should remember that the Robot Commander is designed for educational purposes. Don't expect it to run your household (at least, not yet!). But working within its abilities will give you a good idea of what a humanoid robot can do.

Using Robot Commander with the Computer

The first step after you've built the Commander is to disconnect the motor wires from the wired remote control, and connect the computer (sold separately). You can position the computer anywhere on the robot, but make sure you can still see the display. We found leaning the computer on one of the large feet (without attaching it) worked well, since we could pick it up to enter commands into it, then put it back to run programs.

You'll have to decide which motors you will use with the computer (four can be controlled, while five come with the base Robot Commander set). Two motors can be controlled with a blue T-cable, but they'll work in tandem, and not separately.

We decided to leave the visor motor disconnected for our initial Computer–Commander setup. This gave us two motors for the shoulder and arm, one for the neck swivel, and one for the jaw, yielding the capability for simple automaton behavioral programs. This is how a robot can imitate the actions of a human being.

With the motors connected as above, you can program the Commander to reach and grab a light object from a shelf, while turning its head and turning on its lights to "look" toward the arm that it's controlling. Other actions could include multiple synchronized head and arm motions, with jaw movements. See how much human behavior you can think up for your robot. You may have to make changes to the basic model, such as adding motors. Below are some ideas along this line.

Three-Motor Arm Control

The box for the Robot Commander includes a picture of the shoulder/hand unit with three-motor control (despite the fact that it says you can connect five motors, it shows three). While there are no instructions for this model, it's fairly straightforward to add the motor to the original design based on the picture.

You'll want to do this to give your robot's arm better manipulation and control. The hand in the original design can only open and close.

*The Robot Commander with three
motors in the right arm.*

Adding a motor to rotate the hand will allow the robot to perform actions such as pouring. Make sure you spend some time finding the right materials to work with your robot—a lightweight plastic cup with a large handle and a raised lip should work. Disconnect the motor from the jaw for this project. You'll still be able to program the head to turn toward the work that the robot is doing.

Four–Motor Arm Control

If you want to add four motors to the robot's arm, use the three-motor design discussed above, and add an additional motor to the shoulder (to move the arm up and down). This will concentrate all of the computer's functions in the arm, however, and you won't be able to program any head actions (unless you use a separate computer).

How to Attach Dual Computers for Eight-motor Control

To use two computers with the Robot Commander, we suggest you split the functions. Place four motors in the shoulder, arm, and hand (as described above), to control the arm using Computer 1. Use the second computer for head functions (neck, mouth, and visor).

Program the first computer for the arm routine (lowering the arm, moving it forward, grasping an object, and turning it), and the second computer for the corresponding head actions (turning toward the arm movements, reacting by moving the visor and the jaw, turning on the lights).

You can also use two computers for two-arm control. Either place four motors in each arm (use the right arm as a design guide, and make a mirror copy for the left arm), or put three motors in the right arm plus neck control (for Computer 1) and three motors in the left arm plus jaw or visor control.

With this variation, you could program your robot to perform more complex two-handed tasks. You'll also learn how to fit the design to the task—for example, you may have to put a motor into one arm to make the arm move laterally, instead of up or down. It depends on the task involved.

Additional computers and motors are available from Learning Curve – Robotix.

Further Plans for the Robot Commander

Sitting Figure

You can also reconfigure the Robot Commander to be a seated figure (although the set doesn't include instructions for this idea). That way the arm and hand tasks could be completed with the robot sitting at a table. Reconstruct the robot to make a lap out of the upper leg structure, and rest the robot on the two hip plates. Attach the lower legs at a 90-degree angle to the upper set. Find an adjustable chair with the right width dimensions to hold the hips correctly, and adjust the height to set the feet on the floor.

The seated design could work with the idea of using the *XY* Plotter to create a drawing arm and hand for the Robot Commander. Since the elbow is already bent in the original arm designs, the arm will clear a short table top and provide a base for the drawing arm project.

Controlling the Hip Joint

The torso assembly rotates on a single swivel. This means the robot can be repositioned manually, but the torso–head–arm weight means that you probably can't motorize this component successfully (although four high-speed motors might work).

Motor Drive for the Commander

Another fun project could be creating a set of motorized wheels for the Robot Commander. Attach the motors to the underside of the feet, and use a computer to control them (start with two motors for each foot). You could then use a second computer for arm and head actions.

Be careful if you attempt this project. The swivel joint may be a weak spot for trying to motorize the robot's feet. The weight and torque could cause the top assembly to topple over. We'd suggest rebuilding the torso and hip assembly to be fixed rather than movable before trying to put wheels on the robot's feet. That means taking out the wheels underneath the torso, and attaching the legs directly to it at several points.

Also note that the robot's weight might make this project difficult. But if you can get it to work, your Robot Commander will be able to travel on a route that you program for it, giving it more functionality.

Brain Surgery with the Robot Commander

You might want to think of the Robot Commander kit as a building base for other robotics parts. Since it is currently the only full-scale robot building set available to the public, you could make an efficient hybrid with the robot parts from a smaller kit that has more functionality.

For example, consider that the action of a sensor-based robot computer corresponds to the motors that are connected to it. If you mounted a heavy-duty motor to the hip socket, for example, you could connect it to a robotic computer with a light sensor, and write a program that would make the Commander turn toward a light source. This action would be independent of the Robotix Computer programs, but it would still be a function of the robot.

Further exploration could include wiring a different robot computer to the motors that come with the Commander itself. For example, the Capsela 575 Programmable Control Center runs four motors and has two sensor ports. Using it as a controller for the Robot Commander would mean using a different programming method (see Chapter 2), but you'd be able to link the sensors to the robot's program. Likewise, some of the Basic Stamp controllers discussed in Chapter 7 might work well here as substitute brains.

Note that Capsela and Robotix connectors are not directly compatible (you'll probably have to make your own cables). You should also make sure the voltages for the motors and the battery supplies are compatible. You may also wind up voiding your warranty.

Other Models You Can Build and Control with the Robot Commander Set

The Robot Commander doesn't have instructions for other models (that's left up to your imagination), but there are lots of parts in the kit, and some interesting models are pictured on the box (like a shorter robot figure with a claw hand and wheels, and a large utility vehicle).

We'd suggest getting the Assembly Guide Packs A and B from the Robotix Spare Parts catalog for more ideas. These packs include the instruction manuals for several of the robot kits that Robotix sells, which you can build from standard Robotix parts (like the large number you get with the Robot Commander kit).

Check the parts lists for the other models with the parts in the Robot Commander set, and order the parts not included. In this way, you

An alternate model you can build with the Robot Commander set.

might be able to build Robo-Dog, for example, with the parts from the Robot Commander set.

Also note that the Robot Commander has most of the parts of the Computer Set, so you could just buy the computer and the Robot Commander and still be able to do the projects in the Computer Set manual (included with the stand-alone computer). Check the parts list that comes in the manual to see what extra parts you might need.

Final Thoughts

Robotix balances the concerns of robotics learning with a wide range of innovative building sets. These other sets include a Volcanic Crawler, based on the NASA (National Aeronautics and Space Administration) Dante robot, a Robo-Rex robot dinosaur, and a Robotics Education Set (featuring the computer and a wide range of parts, as well as a Robotics Curriculum).

We'd like to see a continuation of the Robotix line to include an advanced version of the computer. Though you can learn a lot about robotics with the current version, the addition of programmable sensor inputs and a PC interface would make it more flexible.

Contact Information

Robotix kits are available at major toy stores, though you may have to order the computer and the Computer Set specially. Expect to pay around

$100 for the stand-alone computer, $140 for the Computer Set, $120 for the Robo-Dog, and $200 for the Robot Commander. Try the store locator on the Learning Curve web site (www.learningcurve.com) to see where you can get the kits locally.

TurboZ 32.8 SL Programmable Robot Model Car

The TurboZ from SilverLit Electronics is a fully programmable model car. It has a wide range of movements you can program into it via the keypad mounted on top. It's a sophisticated toy that is also a true robot.

The TurboZ programmable robot car.

The TurboZ's programmable keypads.

Features

Inside the nicely designed chassis, a custom microprocessor stores up to sixteen function steps at a time. These range from simple forward, backward, and turning motions to sophisticated preprogrammed routines (sinusoidal curves, irregular polygons, and transverse lines). The eight function keys are used in conjunction with a shift key to program thirty-two distinct movements. Also on the keypad are four duration keys that set the length of the movements.

Three built-in programs show the TurboZ's range of motion. These are accessed using the Demo button and show how well the robot car can zip around.

The TurboZ's internal RAM (random access memory) stores an entered program while the robot is turned on. The current program is erased after a power shutoff, and you can also clear it by using a keyboard command. The TurboZ automatically shuts off after 30 minutes of idle time, saving battery power, and gives an audio warning every 5 minutes that it is idle.

This is a robot car, so it should be no surprise that it makes car noises – like revving up the engine and saying "Let's drive!" in a bright voice every time it starts, as well as screeching through turns and grinding to a stop. These sounds add character to the robot, but be advised that you have to live with them – they can't be turned off (unless you want to disconnect the speaker, which isn't advised).

Programming

Use the keypad to program the TurboZ. Refer to the Command Card for instructions. The commands are in three groups; Beginning Functions, Intermediate Functions, and Advanced Functions.

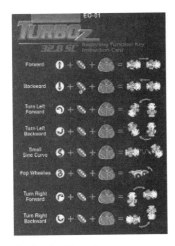

The TurboZ's Beginning Function Key programming card.

Beginning Functions

Forward
Backward
Turn Left Forward
Turn Left Backward
Small Sine Curve
Pop Wheelies (and stay up)
Turn Right Forward
Turn Right Backward

These functions provide coarse control for the TurboZ. For example, the Beginning Function–Turn Left Forward command produces a circular left turn that turns the car around about 1 1/2 times. For more precise robotic control, see the turn commands in the other two sections.

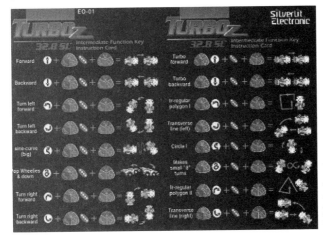

The TurboZ's Intermediate Function Key programming cards.

Intermediate Functions

Forward
Backward
Turn Left Forward
Turn Left Backward
Big Sine Curve
Pop Wheelies (and come back down)
Turn Right Forward
Turn Right Backward
Turbo Forward
Turbo Backward
Irregular Polygon I
Irregular Polygon II
Left Transverse Line
Right Transverse Line
Circle I
Small Figure 8

These functions give you more options for programming an interesting routine into the TurboZ. While the general control is still a bit coarse, you can also add direct subprogram movements (like transverse lines, circles and polygons, and a figure-eight) using only a few commands.

Advanced Functions

Pause
Vibrate
Rotate (Counter-Clockwise)
Rotate (Clockwise)
Shake (three times)
Circle II
Large Figure 8
Pop Wheelies and Shake

Check this section out for the type of control you want for a real robot. The Rotate commands move the TurboZ in generally accurate 45-degree turns, crucial for programming a direct path for the car to follow.

Use a mix of commands to get your robot car up and running. For our purposes, we wanted the TurboZ to be able to follow a discrete path (for example, to make deliveries, or to move through a room without hitting any obstacles). To program a simple square, use the Intermediate Forward command along with the Advanced Rotate Left command, one

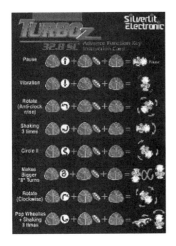

The TurboZ's Advanced Function
Key programming card.

following after the other four times, using the first duration key. A sample program would look like this:

```
Forward
Rotate Left
Forward
Rotate Left
Forward
Rotate Left
Forward
Rotate Left
```

This should give you a good approximation of a square transit. Remember to use the Intermediate Forward command, so your square won't be too large.

Considering that you have to reprogram the TurboZ each time you turn it on, you might want to write down programs that you want to experiment with.

Further Ideas

You can use a programmable car robot like the TurboZ for a variety of tasks.

Robot Delivery

Mount a small delivery basket on the top airfoil. Use the programming keys to set a course from one delivery point to another. Adding the pause command to your program will allow you to stop the TurboZ long enough to have the cargo removed; then it can make its way back to the starting point.

Maze Travel

See how fast you can navigate a maze with the TurboZ by using different programming steps. You may get better control with Intermediate Function steps, but you'll sacrifice speed.

Target Tracking

For this example, set up a flat target ring slightly larger than the TurboZ. Place it five or six feet away from the car (indoors), or even farther away outdoors. Now see who can program the TurboZ to travel as closely inside the ring as possible. Start at the same distance from the target each time.

You can also set up navigation obstacles for this game, and you might also want to do time trials (to see who can get there the fastest).

Wake Up Call

Use the TurboZ's shaking motions at the end of a run to send a wake-up call. Plan the steps to have the car go from one room to another and sound the alarm.

Stunts

The TurboZ has a wide range of motion. You can set up an obstacle course that the car will follow with precision moves. The trick is to place the obstacles (like small traffic cones, etc.) in the areas that the programmed motions will clear.

Set the TurboZ to make a small figure-eight. Mark the starting point and note where the car makes its turns. Place the obstacles in the center of the figure-eight curves. Now when you start the car at the same point, it will appear to clear the obstacles on its own.

Try this with the other interesting program features (like the sine curves, transverse lines, circles, and polygons) to make a full-range stunt course.

Behavior Spoofing

Since you can control the TurboZ through direct programming, you can make it behave in any manner that you want. This can let you simulate intelligent behavior. For example, in the stunt routine above, you could set up a pair of ramps for the robot to jump over. The first one would be at a low height. The second would be obviously too steep and high. Program the TurboZ to go over the first ramp. However, when it gets to the second ramp, have it stop short and shake from side to side (a generally negative motion), and then back up and go around it.

The general observation would be that the TurboZ senses the difference between the two ramps and says "no way" to the second one. That's not what is really happening, but you're simulating that behavior. Try this technique in other situations.

Final Thoughts

The relatively low cost for the TurboZ, combined with its attractive styling (an idea that other companies offering robots could follow), and wide range of programmable moves, more than makes up for its inherent limitations. To push it out of the general toy range, we'd like to see some additional features, like a retained memory and inexpensive programmable sensors (for example, a bumper switch for the front bumper). Another feature that would be helpful would be a PC interface (so that you could save and trade programs that you could download into the TurboZ). We'd also like to see a truck model with a cargo bay in the future.

The SilverLit company has made an interesting foray into the robot market. The TurboZ can be a great introduction to robotics for school-age children up to teenagers and beyond. You can find them at local toy retailers.

Contact Information

TurboZ 32.8 SL (available at local toy stores)
$49.99
SilverLit Toys
P.O. Box 92134
City of Industry, CA 91715
www.turboz.com

Section 2

Intermediate
Robot Kits

Johuco Phoenix

Chapter 6

If you're looking for a small robot kit that comes fully assembled, with features more advanced than those in the previous chapters, you should consider the Johuco Phoenix. It features a strong programming interface (C and assembler) and a PC serial port connection that lets you download and run programs for the robot. Though it is based on a standard remote-control toy car, the modifications by Johuco have transformed it into quite a sophisticated robot for its size and cost. More of an advanced high school or college-level robot, it is worth a look if you want to get into more advanced hands-on robotics and robot programming.

Since the Phoenix comes preassembled, it allows you to get down to the heart of robotics quickly. The twin infrared (IR) sensors allow you to program the robot to follow a wall or find its way out of a maze. The IR and collision sensors work with a preloaded roaming program that will let the Phoenix move about a room freely, avoiding obstacles. Combination programs are provided on the software disk that explore more advanced programming concepts, or you can write your own.

Features

The Phoenix is centered around a small Nikko RC car approximately 6 inches long × 3 1/4 inches wide × 3 inches high, with four rubber tires

The Phoenix robot.

and a standard bottom-mounted battery compartment (for four AAA batteries).

The Phoenix robot circuitry is made up of two circuit boards mounted securely on the base. The MC 68HC11A0 processor is similar to the one in the Rug Warrior Pro™ (see Chapter 12 for more information), and programming it is similar to the Rug Warrior's C programming language.

Two IR sensors are mounted on each side at 45-degree angles. These act as collision detectors, stopping the robot before it actually hits an obstacle. The robot also provides sound and light feedback for these sensors. Two photocells allow you to program light-detection behaviors. Both of these sets of sensors are easily adjustable by use of trim pots (small circular tuning switches) mounted on the circuit boards. A power-level wheel for the IR detection circuits is located near the serial interface in the back.

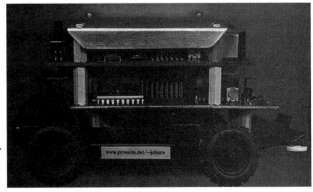

Side view of the Johuco Phoenix.

*Top view of the
Johuco Phoenix.*

A mercury tilt switch that acts as a collision sensor is mounted underneath the main circuit board. This is an interesting device, consisting of a small glass tube filled with liquid mercury and two electrical contacts. When the switch is level, the mercury acts as an electrical conductor. When tilted, or jarred (as when the robot strikes an object), the mercury slides away from the fixed contacts, and the circuit is broken, registering the contact. It's an intriguing, easy way to register a change in a robot's balance.

There's also a small piezoelectric buzzer that you may want to quiet down with a piece of electrical tape, as per the manual.

The serial port interface is a standard RJ11 phone jack connection, with a PC serial cable included. The Phoenix also includes a small expansion interface for adding additional sensors and other devices.

Programming

The Johuco Phoenix has a lot of sophistication inside its small frame. The programming concepts are clearly laid out in a document that comes with the robot, called "Behaviors, Policies, and Activities" (also available at www.pcrealm.net/~johuco/bpa.html). Programming for the Phoenix corresponds to several levels of control. Each level establishes a section of programming code that specifies how the robot will respond.

At the highest level, sequences are grouped as *activities*. These are actions, made up of policies and behaviors, that the robot can switch between during certain predefined conditions.

In the next level, sets of behaviors are collected under *policies*. For example, a drive forward and turn behavior set would be a specific policy. There is some interaction and arbitration between behaviors in a policy, mainly in the way the behaviors are set in the program (the most important behavior is called first, down to the least important).

The basic programming level for the Phoenix is called a *behavior*. The behavior mode controls one actuator at a time (like motor drive and steering). A basic behavior, once defined, can be reused. For example, if you program a turn behavior once, you can reuse it whenever a program needs a turn (as in following a maze). You can also set the time that a behavior will remain active, as in having the robot wait for input from a sensor before performing a behavior.

This is a somewhat complex introduction to programming a robot in a more intuitive way. You can begin to think of the robot as an organism, and of how you'd control and conceive its actions via programming.

The Phoenix has a couple of real-world software interfaces to get the robot up and running, some of which are simple C programming and assembly methods, as well as a way to program in the more conceptual method described above.

To start, use the serial interface cable to make a connection to your PC's serial port. You can then download one of the sample programs included with the software by using a DOS command. Go to the C:\johuco\examples directory and type in DLJ to download a program to the robot.

You may have to modify the initial program to make sure you're connecting to the right serial port. The file dlj.bat (in the \bin directory)

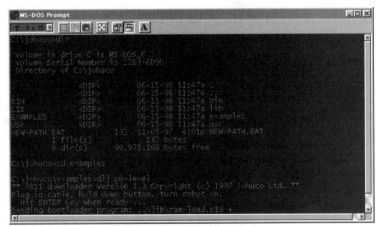

The DOS download window.

has a line for the serial port that can be changed to match your system (under Load11).

The DLJ program loads .S19 program files to the Phoenix. To generate these types of files, use the associated subprograms. For example, to generate an assembly-language program, use the ASJ (ASsemble Johuco) program. This will make the proper .S19 file (and can also download it directly).

Even more useful is the C compiler. Use the CCJ program to assemble a source file and download it to the robot. You can then modify the source file (in an easy-to-edit C format), and change the robot's program.

The default program Wander looks like this in a text editor like WordPad:

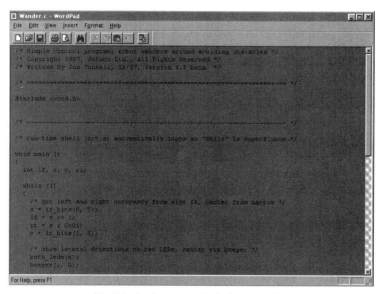

The Wander.C program file, before compiling.

To write your own programs, change the associated variables, and reload the program to the Phoenix. This can get a bit confusing; refer to the file CORD.C (the acronym CORD means Collection of Robot Drivers) in the \libs directory for an idea on what the variables mean.

The Phoenix comes with a freeware version of the ImageCraft C Compiler software, a collection of programs as described above. You may want to check the ImageCraft web site (www.imagecraft.com/software) for an updated version that may be easier to use.

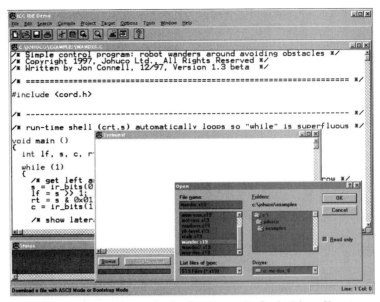

An optional Windows Interactive C software compiler for the Johuco Phoenix.

Sample Programs

Wander

The Wander program, preloaded on the Phoenix, uses the left and right IR emitters/detectors to wander about a room. The robot speeds up when no obstacles are detected, and stops when it gets within range of one. It then beeps, registers the signal on its LED lights, turns and backs up, and continues along a slightly different path to avoid the obstacle. It's a simple avoidance program that can be used for maze-solving and general demonstration purposes. Use the tuning pots on the robot to adjust its performance. This program also uses the mercury tilt switch as a collision sensor.

Wander 2

A modified version of the program above, Wander 2 uses BASAL (BAckground Sensor Actuator Loop) commands, as opposed to free-form C code, to produce a similar behavior for the Phoenix. The BASAL command set is described in the files basal.c and basal.h in the \lib directory.

Try this program to see an example of a basic policy with associated behaviors (as discussed above). The sections in the C file correspond to the behaviors under the specific policy, and they can be used in other, more complex programs.

Stalk

The Stalk program is used to create a specific interesting behavior for the Phoenix. This program uses the light sensors to move to the darkest part of a room, then turns back and starts to ram obstacles at will. The idea is that the robot is using its onboard intelligence to find a hiding place, then attacking from an advantageous position.

The program steps up to the idea of activities. Along with BASAL commands, the activities work with the policies and collections of behaviors to control the robot in a novel way.

Look into the stalk.c file to see this more clearly. There are also instructions on how to modify the parameters of this program to try different activities, policies, and behaviors.

Further Programming Suggestions

The Johuco Phoenix manual has a section entitled "Software Ideas." Use the suggested methods to try out variant behaviors for your robot.

Follow the Remote

This routine will allow the Phoenix to home in on a television remote control by using the IR sensors. It's an intriguing alternative control system.

Find the Exit

By using the photosensors, you can have the Phoenix find bright areas in the room. The general idea is to have the robot turn to the left, determine if the light source falling on it is brighter, and turn back to the right if it's not. Then the robot should turn to the right and repeat the process. This way, the Phoenix should work its way out of a dark room toward a lit doorway.

Dr. Jekyll and Mr. Hyde

The placement of the photosensors (one in front and one on top) means that your Phoenix can tell the difference between "day" and "night." Write a program that uses the top light sensor in this manner. In the daylight, the robot will run the standard Wander program. At night, the robot could change personality and run a ramming program, striking obstacles at will. You can even change the robot's personality by switching the lights off and on.

Final Thoughts

The battery upgrade, as suggested in the manual, is a good idea. The flat metal base makes an ideal spot for a bigger battery pack. You may also want to consider placing a small basket on top of the robot to allow it to carry cargo.

Use the base of the Phoenix to make a robot from a larger RC car. You'll need to match the voltages that the robot uses for its central processing unit (CPU), and make sure that you get the same general kind of remote-control (RC) car, one with rear wheel drive and front wheel steering.

At $299, this small robot may be a bit expensive for some. But the high degree of skill that goes into the pre-assembled Phoenix means that there is a shorter electronics learning curve when you get started.

The main concern is that the Phoenix leaves a lot of the programming essentials up to you. You should have a good idea of how a structured programming language like C works to get the full benefit from this robot, and you'll also have to look deeply into the concepts behind the innovative behaviors, policies, and activities control schema to get the full benefit from them.

There are many books available on C programming, from publishers like Addison-Wesley (www.awl-he.com), SAMS (www.mcp.com/publishers/sams), and Microsoft Press (www.microsoft.com/mspress). You can also look into C software and tutorials from vendors like Borland (www.borland.com) and Microsoft (www.microsoft.com).

Contact Information

Phoenix Mobile Robot
$299, fully assembled
Johuco, Ltd.
Box 385
Vernon, CT 06066
www.pcrealm.net/~johuco

Chapter 7

StampBug II

T his robot, StampBug II, from Parallax, Inc., and Milford Electronics, is a strong entry in the low-cost robotics field. It emulates the walking motion of an insect and can be programmed to change directions when it senses obstacles with sensor feelers. The kit also lets you get involved with Basic Stamp microcontrollers and programming, a low-cost hardware and software electronics combination with many useful applications, including robust robotics development.

The StampBug II.

The Basic Stamp is an innovative microprocessor that takes up very little power and space. It runs a modified version of the BASIC programming language, and it has a wide variety of uses for creating and controlling small electronic devices. It also makes a good robot controller.

The Stamp Bug II kit features a Basic Stamp controller, three Hitec servos to control movement, and all the parts you need to make a sturdy battery-powered six-legged walker robot. The kit also includes two touch sensors that you mount on the front of the bug to act as feelers, and BASIC software to download programs that you create (a sample program for the Stamp Bug is also included).

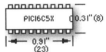

The BASIC Stamp command interpreter chip.

A fair amount of soldering and mechanical assembly is required to put this kit together properly, but it's a solid performer once you're done.

The bug walks using the servos to move the front and back legs forward, canting the middle leg on one side to pivot it up and forward in the same direction. It then repeats that action on the opposite side, propelling itself along. It is an innovative design that shows what can be done with only three servos.

There's a jumper switch that allows you to cast the middle leg farther down, to allow the Stamp Bug to move over different terrain (like a deep-pile carpet).

The two sensor feelers are used in the example program as obstacle detectors. These are slender metal wires that make contact with the sides of metal sockets to register obstacles. The program makes the Stamp Bug walk about freely until one of the sensor feelers hits an object. The Bug will then back up, and turn in the opposite direction. In this way, the Stamp Bug can wander about a room, and eventually find an exit.

A bottom view of the StampBug.

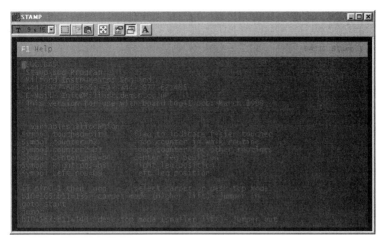

The STAMP.EXE BASIC editing program with the bug.bas control program loaded into it.

Programming

You program the Stamp Bug (and other Basic Stamp robots) in BASIC. The example file is called bug.bas, and consists of a text file that is loaded into the STAMP.EXE application.

You can edit it in that application and download it to the Stamp Bug via an included serial cable. The bug.bas file can also be edited in a word processor, but make sure you save it in the original ASCII text format.

Note that the variable editing is a lot like the C files for a robot like the Rug Warrior Pro™ (see Chapter 12). Change the relevant sections to change the way your Stamp Bug moves around. To make it follow a path, write a program with forward motion and left and right turns. This type of navigation is known as dead reckoning. It's the same type used for other robots in this book. Make sure you plot your course before you start programming the robot, and experiment with the robot's range of motion (for example, how much rotation is needed to complete a 45-degree left turn).

Further Ideas

With the basic kit, see if you can change the programming to have the bug follow a wall or work its way through a maze. Change the relevant variables to make sure that the Stamp Bug II doesn't get stuck in corners, and see how alternate programming values for movement affect the Bug's range of motion.

You should also check the Parallax web site for a lot of ideas for experimenting with your Stamp Bug II (www.parallaxinc.com/stamps/ stampbug.htm). These include things you can do with the basic kit, like changing the stride lengths of your walker's movement (to adapt it to different terrains), and adding visual feedback during maneuvering (i.e., blinking the front LEDs when the Bug backs up). There's also information on how to add different sensors (like photodetectors, to sense obstacles), and a lot of technical notes.

Other Sources of Basic Stamp Robotics Information

There's a wide range of Basic Stamp information on the web. The Parallax web site has a Links page with some good starting points: http:// www.parallaxinc.com/links/links.htm

Final Thoughts

The Basic Stamp is an excellent microcontroller for robotics. It's important to be able to start with a good low-cost controller in a robot kit like the Stamp Bug. You learn how a robot depends on its control system to give it functionality.

Having said that, it would be interesting to see a user interface beyond BASIC for the Basic Stamp when used for robotics. A graphical control system goes a long way toward making robots easier to use.

Expect Parallax to keep developing robot kits. In the works is a Basic Stamp-based kit called the GrowBot, which should feature a wheeled base like the Rug Warrior Pro™, and expansion options such as various sensors and related devices.

Contact Information

StampBug II
$139.00
Parallax
3805 Atherton Road #102
Rocklin, CA 95765
916-624-8333
www.parallaxinc.com

BEAM Robotics

B EAM robotics is an alternate method of constructing robots that's currently gaining in popularity. Developed by robotics researcher Mark Tilden, the acronym BEAM stands for Biology, Electronics, Aesthetics, and Mechanics. The general idea is to use biological ideas to make robots, using electronics to mimic living systems through innovative kinematics (mechanics), and adding aesthetics to the mix.

For example, instead of the traditional system of programmable logic connected to a battery or other source of power, some BEAM robots are self-contained units, drawing power from the sun. They collect the solar energy in capacitors, storing electricity until they receive enough to discharge it to a motor. Just like any other robot, you can add sensors to allow a BEAM robot to react to its surroundings.

Other BEAM robots use battery power but use biological models for their circuit systems. Types of BEAM robots that have been developed include small walkers, and the scuttling bugs described below.

The PhotoPopper

The PhotoPopper Photovore kit, from Solarbotics (Calgary, Alberta, Canada), is a great example of BEAM robotics in action. It's an ad-

The PhotoPopper
Photovore.

vanced kit that requires good soldering and electromechanical assembly skills, but the instructions are clear, and the results are worth the effort you put into it.

The PhotoPopper is a small, insect-sized robot kit, about 3 1/2 inches long × 3 inches wide × 1 inch high. The completed model features a solar cell to supply power, a capacitor to store electricity, and two pager motors for movement. The sensor array consists of two touch sensor whiskers and two photocell light sensors.

To activate the robot, simply place it under a strong light. There's no on/off switch, and you will need to put it into a dark box to render it dormant. The PhotoPopper has been designed around a light-following program that works well with its overall construction. The robot will automatically turn toward the brightest light source, using the light sensors (mounted left and right) to make this judgment. It will then direct the power to the opposite side motor to turn the robot toward the light.

The touch sensors work in somewhat the same way. If one feeler hits an obstacle, the PhotoPopper will turn away from that side, allowing the robot to go around it.

Programming the Photovore

Note that you can't enter a program into the Photovore directly. The only way to change the routine for the robot is to change its environment. For example, you can control the movement of the Photovore in a dark room by moving a strong light around it. You could also use lights on a timer to move the PhotoPopper along a path. By turning lights on and off, you would "pull" the robot in different directions.

The PhotoPopper can also work as a maze-follower, using its touch sensors to avoid obstacles. For this scheme to work, you should put a strong uniform light above the far end of the maze, so that the Photovore will move toward it, and make sure that the solar cell doesn't get trapped in shadow.

Final Thoughts – PhotoPopper

The PhotoPopper is a stimulating advanced-level kit. You'll need good soldering skills to get the Photovore put together properly, but once you have it up and running, you'll appreciate the clever design. The price is right, too, and you'll never need batteries!

Contact Information

PhotoPopper Photovore
$60
Solarbotics, Inc.
179 Harvest Glen Way N.E.
Calgary, Alberta, Canada
T3K 4J4
(403) 818-3374; (403) 226-3793
www.solarbotics.com

Also available from:

Mondo-Tronics' Robot Store
4286 Redwood Hwy, PMB-N #226
San Rafael, CA 94903
(415) 491-4600
www.robotstore.com or www.Mondo.com

CYBUG

Another great BEAM kit is the CyBug. This kit creates a larger robot bug that demonstrates some interesting behaviors.

The basic CyBug is powered by a 9-volt rechargeable battery and has two small direct-current (DC) motors that drive the robot about like the Photovore (but the CyBug's are larger and faster). The behavior is controlled by two photocell light sensors and two bump switches that act as fenders.

There's a small jumper setting at the top of the CyBug that switches it between light-following and light-avoiding behaviors. In the first case, the CyBug will follow a light source, making course corrections depending on whether it is picking up more light on the left or the right photocell. You can control the CyBug's direction by slowly moving a light around it in a dark room. In light-avoiding mode, the robot will avoid a light source,

The CyBug.

scuttling away from the side that gets the most light (a photophobic response that closely mimics the behavior of a wide range of insects).

Programming The CyBug

You control the CyBug's behavior by using a potentiometer to set a fast (aggressive) or slow (timid) response time. The bump switch sensors are a part of a built-in subprogram that keeps the CyBug from getting stuck in corners and helps it to get past obstacles. The general idea is that if the fender on the left side registers a hit, the bug will back up and turn to the right, and vice versa.

Use the potentiometer system to tune your CyBug, for navigating a maze, for example. Figuring out the proper response time to a bumper hit could help it to work its way around corners without getting stuck.

Add-ons for the CyBug

Hunger Instinct

The HBF-1 add-on is a daughterboard that programs your CyBug with a feeding instinct. It sits on top of the CyBug and monitors the rechargeable battery supply. The general behavior for this CyBug is to avoid a light source, scuttling away from the brightest part of a room. But what happens when the battery supply gets low? Fortunately, this robot can figure out how to refuel itself.

When the battery voltage drops below 7 volts, the robot becomes light-seeking and will head off to a feeding station you can construct for it. This station is usually a small box with a light mounted on the inside. In a dark room, it becomes the target for the modified CyBug. The inside

*The CyBug with the
Hunger Instinct add-on.*

top and bottom of the box are covered with tinfoil, and they act as electrical conductors, touching little feeler antennas on the CyBug and sending power to the robot's battery.

As the robot is charged, the HBF-1 monitors the voltage. When the charge is high enough, it changes the CyBug's behavior back to the general light-avoidance pattern, and the robot zooms away from the changing station until the next time it needs to feed itself. It's an impressive display of robotic functionality.

Predator/Prey

The HBF-2 daughterboards provide an intriguing interaction between two CyBugs. You mount the first board on the CyBug that will be the predator. It features two IR sensor "eyes" that will control the robot's movements.

The prey daughterboard goes on another basic CyBug. This board features an IR-emitting diode that sends a signal out from the back of the prey bug.

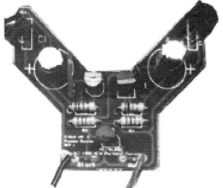

The CyBug Predator Board.

The CyBug Prey Board.

Now the two reprogrammed bugs will exhibit a classic chase–flight pattern. The prey bug gives off an IR signal that the predator can detect. The predator's IR eyes are mounted on left and right sides of the robot and work with the motors to change the robot's direction (depending on which side gets the strongest IR signal). This keeps the predator on track as the prey scuttles away from it.

Final Thoughts - CyBug

The CyBug is a neat application of the BEAM theory of robotics. It bypasses a lot of the preconceptions about robots that have kept them out of the hands of the general public, and presents remarkable robot behavior at a great price.

Contact Information

Cybug Kit, $49
Hunger Instinct HBF-1 add-on, $21
Predator-Prey HBF-2 add-on, $25
CyBug Space,
http://members.home.net:80/cybug/

Also available from:

Mondo-tronics' Robot Store
4286 Redwood Hwy, PMB-N #226
San Rafael, CA 94903
(415) 491-4600
www.robotstore.com

HVW Technologies
Suite 473, 300-8120 Beddington Blvd. N.W.
Calgary, Alberta T3K 2A8
Canada
(403) 730-8603
www.hvwtech.com

CANTronics
440-10816 MacLeod Trail S., Suite 233
Calgary, Alberta, T2J-5N8
Canada
cantronics.rzsoft.com

Technology Education Index
40 Wellington Road, Orpington, Kent, BR5 4AQ
United Kingdom
Within the UK: (01689) 876880
Outside the UK: (+44) 1689 876880
www.technologyindex.com

Robotics World
100 Jalan Sultan, #02-16 Sultan Plaza
Singapore 199001
(65) 298-8426

Final Thoughts

BEAM robotics, on the whole, seems like a good idea. It provides an entry point for designing low-cost robots. The general principle of using biological models to construct complex robots using streamlined techniques is sound.

We'd like to see the aesthetic principle that's supposedly a part of the BEAM philosophy given a bit more consideration. Most of these kits (indeed, this could be said of the majority of the products of the entire robot industry) still look like the inside of a machine, rather than the outside. We could imagine a horse that simply had an skeleton with bunches of muscle attached to it, but that doesn't happen in nature. The car companies wouldn't try to sell you a car without a car body, even though it would still run. It would be a great improvement for the BEAM bugs to incorporate the fascinating exterior designs that nature has developed over the aeons (like a stag beetle carapace, for example). More attention to design has to develop.

*An Albrecht Dürer
stag beetle.*

Also, the break from previous traditional robot designs means that you have to understand the benefits and limitations of BEAM robots before you begin. Since they're generally more like living organisms, they have programming built inside them.

You can usually change the programming either by mechanical switches and jumpers, or by adding additional electronics, but note that you can't modify the robot's behavior by editing a downloadable control program file. For example, you can't program one of the kit robots described above for dead-reckoning navigation (that is, to have the robot travel along a repeatable exact path). This issue could be cleared up in subsequent BEAM robots. We'd like to see a BASIC Stamp model in the future; the combination could allow the BEAM robots to access the more complex interactions with sensors that traditional programming allows, and would make them more fully programmable.

BEAM robotics continues to be interesting. Check the web sites for Solarbotics (www.solarbotics.com) and CyBug Space (http://members.home.net:80/cybug/) for updated kits and links to more information.

OWI Elekits

O WI Elekit robots provide an excellent tutorial for robotics principles. Of special interest are two kits that use sensor-based navigation and programming (The WAO II and Navius robots). Each robot is easy to construct, and the kits are very well manufactured. Parts are color-coded, and the instructions are comprehensive.

The WAO II kit has programmable sensors and motion control (through an on-board keypad or downloadable from a PC using MS-BASIC), two-motor wheeled drive, and a piezo-electric buzzer. You can also attach a pen to the bottom of the robot for a limited drawing function linked directly to the drive system.

Each robot is a self-contained package that will give you a good understanding of the principles it illustrates.

Navius

The Navius robot from OWI Elekit is an interesting example of how to program a robotic device by a hard-copy input method. The base unit is approximately 2 3/4 inches high × 6 1/4 inches wide × 5 inches high and consists of a small platform with two clear domes (one for the electronic

The Navius robot.

control system and one for the programmable disk and sensor). It has an on-board navigation sensor that uses user-programmable disks to control the motion of two drive wheels. Three motors beneath the platform are powered by a pair of AA batteries located under the sensor dome area (one set for the wheels, and one set to move the program disk past the sensors). A 9-volt battery under the electronics area powers the control system. Battery life is estimated at 60 minutes continuous or two hours intermittent, with alkaline batteries installed. The weight is distributed evenly throughout, and the unit is supported by a rear caster.

Theory of Operation

Navius uses a removable, programmable paper disk with encoded marks that is read by its on-board IR sensor system. The sensor information is collected, converted to electric impulses decoded by the logic circuit, and passed to the drive system. The Navius unit reads the program disk to find out which motions it should make, much as a human reads a map. Its internal sensors tell it how many steps to take, when to turn, etc., making it possible to move about the real world and follow a maze or avoid obstacles (as long as the programming accurately reflects the environment).

The sensors are arranged by pair for each of the two motors. They bounce an IR beam off the surface of the disk as it spins, and convert the reflected signal into electric impulses. A white space on the disk is reflected as a high signal, and a black area is reflected as a low signal. Low signals from the black marks cause the motors to turn. Each disk has two channels, one for each motor. By making both channels black, the programmer can make Navius move in a straight line. Turns are achieved by marking one channel black and the other white.

*A closer view of the
Navius robot.*

The IC4584 on the main logic board takes the electric signals from the infrared sensors and outputs the proper voltage (high current for no or a low signal, and no current for a high signal). The circuit ignores small fluctuations in ambient light.

Navius is a true example of binary coding, the foundation of computer programming. Each signal is seen as a high or a low current, and is treated correspondingly. The programming is branched out by the amplifier, which provides power to the motors when the signal is high. Two photoresistors read the signals for the left and right motors independently.

Note that the surface that Navius is on (smooth ones work best), the amount of power in the batteries, and the physical meshing of the gears (are they greased properly? are foreign materials present?) affect performance.

Programming Navius

To program Navius, take a blank program disk and black out the inside, outside, or both sectors for right, left, and forward movement. Use drawing paper for the best effect, and a dark pencil or black ink. Cut out the blank disks included with Navius to get started. You may want to make high-quality copies of the disk page before you start cutting. This is an easy way to make more disks. You can also use the Card Gear Panel and a compass or protractor to draw more disks. There are three small holes on the Gear Panel (located directly under the program disk) you can use to draw the circle lines. It's a bit tricky this way, so make sure you have backup blank disks on hand.

Navius reads the input disk at a steady rate of 21 seconds per revolution (0.58 seconds per sector, 36 sectors per disk). It's important to match the programming to the time that the disk travels across the sensor.

Navius is also limited to repeating the pattern on the disk ad infinitem as the program wheel rotates continuously when the robot is activated. Adjust your programs accordingly. Use more sectors for a wide turn, less for a short turn. Practice by making a short route with one or two turns that Navius can follow, and a continuous loop. You may have to tune the sensors carefully to get the proper response (see the manual for more information).

Motion Control for Navius

```
Right turn (left motor drives)- black along outside
    track
Left turn (right motor drives)- black along inside
    track
Forward (left and right motor drives)- black along
    both tracks
Stop - white on both tracks
```

Experiment with the unit to see how to get the best motion effects.

Documentation

The Navius manual is a wonderful construction guide, beautifully illustrated, but it leaves the discussion of the operation of the robot a bit short. There is also a good deal of technical information, including circuit diagrams, electronic circuit block diagrams and descriptions, and an entire section on gear functions and torque effects, and how they relate to wheel speed and program disk speed. It may seem a bit bewildering, but the in-depth knowledge of how a real robot is designed (how the power system relates to the control system, for example, and how the gear system creates enough torque to move the robot about) is essential to an understanding of robotic principles. You'll find that as you move to larger robot kits and your own projects, the knowledge will come in handy.

Special Projects

It should be possible to adapt the sensor assembly to make Navius a bit more flexible. For example, a tape recorder assembly would provide the starting point. The idea is to replace the disk with an alternate input source. Using a paper tape that rolled past the sensor assembly (with the proper markings), a modified Navius unit would be able to run longer programs. The main caveat is that the sensor assembly on Navius is a bit wide. This wouldn't be a project for beginners, but it's a possibility.

Final Thoughts

Navius is an interesting example of a robot controlled by a hard-copy program. Future robots with this principle could use easy-to-make command cards (or disks) for specific tasks. For example, a household robot could have cards that told it the way to get the cat food and deliver it to the proper bowl, and the best route to pick up and take out the trash. If you move the trash can, it would be a simple matter to make a new card for the robot showing the new route. This eliminates the need to run a computer program, or mess about with the robot's internal settings.

Note that this is a relatively simple robot, and it has no external sensors. But it does give a good example of a flexible programming method that works.

WAO II

The WAO II is a programmable robot that uses a 4-bit microprocessor to store and execute commands. The unit is approximately 5 1/2 inches wide and 4 1/2 inches tall, in a round plastic case with a clear dome that allows you to see the electronic components. Two wire feelers connect to left and right sensors that can be programmed independently. The unit rests on two wheels and a central caster. There's also a penholder on the undercarriage that you can use for some limited drawing exercises linked to the drive system.

The kit is easy to build. The parts are well laid out, and different sections of the robot are molded in different colors (white for the external case, green for the chassis and undercarriage).

You can give the WAO II direct commands via the keyboard that makes up the back of the robot. You also use this keypad to enter a series

A closeup of the WAO II robot.

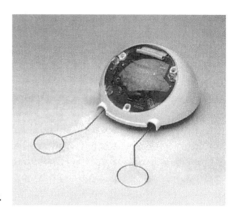

The WAO II robot.

of commands to store in the system memory and run as a program. An accessory cable for transferring programs to the robot from a PC is also available.

The unit runs off a 9-volt battery for the logic circuit and three AA batteries for the motor drive system. Battery life is estimated at two hours for the 9-volt battery and 20 minutes for the AAs with continuous motion.

The WAO II uses a special sleep routine to conserve power, corresponding to the Run/Sleep key on the main keypad. The robot will still retain the program in memory, but the power used will drop considerably. For extra power conservation, the unit will go into sleep mode automatically after two minutes of inactivity. WAO II will also go into sleep mode after running a program or if the Run/Sleep key is pressed.

WAO II's onboard keypad has commands for direct demonstration programs stored in read-only memory (ROM), movement keys that work in direct and program modes, and special program-only keys (printed in orange). The keypad also has a system status LED that indicates the mode that the robot is operating under, a Reset switch, and the Run/Sleep key.

There are a total of twenty-six keys on the pad, and you can enter up to twenty-four commands (or command sets) at one time. Entering more than twenty-four steps will send the WAO II into sleep mode by overloading it with information.

WAO II also includes a programmable piezoelectric buzzer that sounds a peep when activated. The buzzer is a small black circular object located in the lower front center of the robot, under the clear dome. It is also used to confirm entered commands (one peep), and to report errors (two peeps for incorrectly entered commands, five peeps for errors in a stored program).

Picture of keypad.

Theory of Operation

The WAO II uses a 4-bit microprocessor to store and run programs. The integrated circuit (IC) uses ROM to correspond to action commands entered into the robot from the keypad, or from a program stored in RAM. The robot can be configured to use conditional action, a key step in the path to an intelligent machine. IF statements correspond to decisions that the robot makes as it navigates about the real world.

You can also change the way that WAO II reacts to sensor input for different scenarios, like using a specific memory to perform different tasks. For instance, the robot can be programmed to avoid walls by monitoring its left and right sensors for a sensor hit (the feeler is pushed into the robot and triggers a microswitch), and taking the appropriate action. However, the same program would not work in a different scenario, like cliff avoidance. You would enter an alternate program for that situation, and change the way the robot monitors its sensors. In that case, the wire feelers would be angled down, pushing the sensor switches on, and the robot would monitor them for an off state that would signal a drop.

See the programming section that follows for examples of both of these routines.

WAO II Direct-Mode Control

Demonstration Programs in Direct Mode

Push the Reset key, and then the Run/Sleep key, to set WAO II into direct mode. You can then run the demo programs from the onboard control panel.

Roulette

After pressing the Roulette key, the WAO II will turn in a random direction, gradually slowing to a stop. A built-in program runs a random sequence to the robot's motors. You can make a circular drawing with colored sections like a roulette wheel (use approximately eight sections, and make the circle 10 inches or more in diameter for the best effect). WAO II can then be used to pick a number at random, or for a fortune-telling game (put simple answers in each section, like Yes–No–Maybe–Ask Again, and so forth).

Dice

Using the Dice key in Direct Mode will cause the WAO II to draw a number from 1 to 6. Put a pen filler into the holder underneath the robot and adjust it carefully (see the manual for more information). Use a wide piece of paper, or tape several 8 1/2 × 11 inch sheets together. Press the Dice key, and WAO II will draw a number from 1 to 6, using the direct motions of the wheels to draw the number.

This is an example of a stored program that randomly picks one of a set of six motions that correspond to the "numbers", and executes the actions. Make sure you get the pen adjusted properly, since it has to rest lightly touching the surface of the paper, and it can interfere with the wheels if it is extended too far (or not draw at all if it is too short).

The WAO II uses broad curves and sharp lines to draw the numbers, and the results can be hard to read. You may have to spend some time adjusting things to get this feature to work properly. Note that the numbers are of irregular sizes, and that this is the only example of the WAO II's drawing ability. Unfortunately, the exact commands used to draw the numbers are not specified. You may be able to interpret the commands by watching the WAO II at work.

See if you can build up a set of programs that correspond with each number. Otherwise, use the Dice program to draw a number at random.

Timer

The Timer key sets the WAO II for delayed action. Press the Timer key and then the 1 key, and the WAO II will wait for approximately 1 minute, then turn to the right once. Use the other number keys to increase the number of minutes that WAO II waits and the corresponding number of turns (for example, Timer plus the 9 key will cause WAO II to wait 9 minutes, then turn to the right nine times). This illustrates the use of a delay command to schedule a distinct program at a specific time.

Direct Input Commands

You can use the motion controls on the keyboard to move the WAO II in direct mode. For example, use the forward arrow to move forward, and a number key to set the duration of the movement. You can also initiate direct movement with the Turn, Pivot, and Stop keys, as well as the piezoelectric buzzer. Experiment with the movement keys in direct mode to learn more about WAO II's range of motion. Use the direct keys with the writing system to step through a drawing sequentially. Just remember to write down the steps, so you can write a program to run through them all at once.

Note that the WAO II waits 1 second before moving, and that each number key is a multiple of 250 milliseconds (1/4 of a second). So, for example, pressing the forward arrow key and the 2 key will result in 1/2 second of forward movement. Circular motion is set at 125 milliseconds (1/8 of a second).

Programming the WAO II

The WAO II is able to store a program in memory. This is a set of steps like those found in direct-mode control. By stringing steps together, you can make WAO II do more complex tasks. Note that the WAO II can only store a single program, and that the Reset button clears the robot's memory. You can, however, put the robot into sleep mode and still retain the program in memory.

You can also use programming to set conditions for the WAO II sensors, and combine motion and sensing to get the robot to do some cool routines – by itself!

Programming with the Onboard Keypad

A simple program that illustrates WAO II control follows.

Push the Reset key to set WAO II to program mode. You can enter a maximum of twenty-four program commands and number value sets.

Square

```
Forward arrow + 2
Stop + 1
Right pivot + 4
Stop + 1
Forward arrow + 2
Stop + 1
Right pivot + 4
```

```
Stop + 1
Forward arrow + 2
Stop + 1
Right pivot + 4
Stop + 1
Forward arrow + 2
Stop + 1
Right pivot + 4
Stop + 1
```

Note that this amounts to four sets of commands repeated four times (one for each side of the square), or sixteen programming commands. You may have to adjust the commands to get WAO II to draw a proper square.

It can be a little tedious to input similar commands, so WAO II also includes special programming codes for repeat functions (FOR and NEXT). To draw the same square as above, hit the Reset key and enter the following:

```
FOR + 4
Forward arrow + 2
STOP + 1
Right pivot + 4
STOP + 1
NEXT
```

The FOR key plus a number value equals the time the routine will be repeated. The routine ends with the NEXT statement. The above example puts the side of the square-drawing routine that we originally repeated four times into a FOR–NEXT loop, and runs it four times to get the proper result.

Note that the FOR + 9 command set is for infinity, and not for nine times. This is a useful programming command that allows the WAO II to have continuous conditional motion. Here's an example of a square-drawing routine that repeats continuously:

```
FOR + 9 (infinity)
FOR + 4
Forward arrow + 2
STOP + 1
Right pivot + 4
STOP + 1
NEXT
Right pivot + 1
STOP + 1
NEXT
```

There are two sets of FOR–NEXT commands, one inside the other. The inside set draws the square with the familiar commands. The outside set repeats the inside set continuously, and adds a Right pivot + 1 to stagger the squares (so that the WAO II doesn't draw them on top of each other).

See what kinds of geometric drawings you can make with these command sets. Consider a polygon, for example. Use the FOR–NEXT commands to repeat the geometric shapes to make more complex drawings. Here's a circle-drawing program that will draw slightly offset rings to make a circular pattern.

Circle

```
FOR + 9 (infinity)
FOR + 2
Left turn + 8
NEXT
Left pivot + 1
STOP + 1
NEXT
```

Programming the Sensors

The WAO II comes with two tactile sensors, one on the left and one on the right. These are microswitches that are monitored by the robot. The "feelers" can be set to activate specific programming under specific conditions. Special programming keys control the sensors' state. On the robot's control panel, the IF and ENDIF keys are used with the SL ON, SR ON, SL OFF, and SR OFF keys (numbers 1–4 on the number pad) to add sensor functions to programs. The IF command starts the conditional state, and the ENDIF command closes the routine. The SL and SR commands correspond to the left sensor and right sensor "on" and "off" states. The sensor is on when the switch is pressed down and is off when it isn't. Two wire antennas extend from the switches and can be positioned to give interesting effects from the sensors.

By programming the sensors, you can set WAO II to react to a sensor "hit" with a conditional program. When the robot receives the information that matches the condition, it runs the program associated with it. The analogy is like moving your hand forward until you strike a wall, and then moving it back to avoid hitting it repeatedly. The brain receives the signal from the touch receptors in your fingers and makes a judgment that another direction of movement is needed.

Here's a sample program that allows WAO II to follow a wall without jamming up against it. The IF–ENDIF commands allow the right and left sensors to activate fending-off programs that move the robot away from an obstacle. They also allow forward movement when no sensor hit is registered for either sensor.

```
FOR + 9 (infinity)
IF + 3 (SL OFF)
IF + 4 (SR OFF)
Forward arrow + 9
END IF
END IF
IF + 1 (SL ON)
Reverse arrow
Right pivot + 3
END IF
IF  + 2 (SR ON)
Reverse arrow + 2
Left pivot + 3
END IF
NEXT
```

To break this down into normal language, this program tells the WAO II to move forward continuously, as long as it doesn't feel any obstacles (both sensors are off). If the left sensor touches a wall, the robot will back up and turn to the right before moving forward again. If the right sensor touches a wall, WAO II backs up and turns to the left before continuing. The robot is seeking a clear path to travel, and adjusts its direction accordingly.

This is a great demonstration of a robot in action. This simple program will allow WAO II to wander continuously around a room without getting stuck (although you may have to make sure that the sensors don't get snagged and that the robot doesn't run into any low obstacles, like a carpet).

Maze Following with Right-Avoidance

The WAO II can find its way through a maze by using wall-sensing programs. The analogy would be that you tapped along a wall in a dark room to find an exit. Each time your fingers hit the wall, you would back up and move forward until you touched it again, repeating the process until you found your way out.

```
IF + 9 (infinity)
IF + 4 (SR OFF)
Right turn + 1
IF + 2 (SR ON)
Left pivot + 2
END IF
END IF
```

```
IF + 4 (SR OFF)
FORWARD + 3
END IF
IF + 2 (SR ON)
Reverse arrow + 1
Left pivot + 2
END IF
NEXT
```

Set up a maze with walls that are at least 6 inches high and corridors about 12 inches wide. Remove the left feeler; you won't need it for this program.

The WAO II is able to navigate a simple maze with this program. The right turn keeps the robot trailing along the wall. The forward motion moves the robot along, unless the right sensor registers a hit (by touching the wall). Then the robot will back up by pivoting to the left, turn off the right sensor, and continue the primary actions until the exit is found. You can also rewrite the program to sense on the left by changing some of the parameters and switching the sensor to the other side.

Cliff Avoidance

To try a different approach, see how the robot can scan for sensors that should be constantly "on", and take appropriate actions if they aren't. In this example, the WAO II avoids falling off a table by feeling for the table surface with its sensors. If one sensor registers "off", the "drop" is sensed, and the robot takes appropriate action, backing up and turning in the other direction before continuing. Set the sensors at a 45-degree angle downward, and be careful that the robot doesn't fall off the table during backward movement.

```
FOR + 9 (infinity)
IF + 1 (SL ON)
IF + 2 (SR ON)
Forward arrow + 1
STOP + 1
END IF
END IF
IF + 3 (SL OFF)
Reverse arrow + 2
Left pivot + 3
END IF
IF + 4 (SR OFF)
Reverse arrow + 2
Right pivot + 3
END IF
NEXT
```

Now try a program that will allow WAO II to stay *under* a table by feeling for the underside with its sensors. This could be used to keep a robot under cover during rainy weather, for example. Note that this will require constructing a platform that the WAO II's sensors will reach, and that you'll have to make modified feelers as well, since the originals can't be set to reach over WAO II's head. OWI sells replacement feelers for a small fee; see the manual for more information. Don't use the original sensor feelers that came with the WAO II – bending them may cause them to break off.

To make the feelers for this program: Bend the feeler up at a perpendicular 90-degree angle about one inch from the base of the feeler wire. Attach it to the holder, and connect it to the sensor at a 45-degree angle. It should clear the top of the WAO II.

To make the table: The table should be about 5 inches tall and level. It should be heavy enough to keep the sensors pressed down. To get the proper height, measure the modified sensors in place when engaged (pressed down). Adjust the sensors to fit the table. You need enough clearance to allow the WAO II to move about, but not so much that the sensors don't feel the underside of the table. Make the table any width and length you like.

The program for keeping the robot under a table is the same as the one for ensuring that the robot avoids a fall. You're just switching the sensors about to give the WAO II a new capability.

```
FOR + 9 (infinity)
IF + 1 (SL ON)
IF + 2 (SR ON)
Forward arrow + 1
STOP + 1
END IF
END IF
IF + 3 (SL OFF)
Reverse arrow + 2
Left pivot + 3
END IF
IF + 4 (SR OFF)
Reverse arrow + 2
Right pivot + 3
END IF
NEXT
```

Now the WAO II will approach the end of the table, sense that the underside is not there, and dart back under in a reverse motion. It will repeat these actions, staying under the table indefinitely.

Combining Sensor Action

You can combine different sensor actions to make the WAO II do interesting tasks. In this example, the robot will move forward and probe for an exit on the ride side of a wall, while avoiding a cliff on its left side (by sensing the ground).

You'll have to construct the platform for this example. Make a ledge about 1 foot off the ground and a foot wide, with a 6-inch wall on the right side. The wall can be as long as you like. At one end, make an exit hole about a foot wide. Angle the left sensor 45 degrees downward and the right sensor 45 degrees up.

Use a program like this one:

```
FOR + 9 (infinity)
IF + 1 (SL ON)
IF + 4 (SR OFF)
Right turn + 1
IF + 2 (SR ON)
Left pivot + 1
IF + 3 (SL OFF)
Right turn + 1
END IF
END IF
END IF
END IF
IF + 4 (SR OFF)
Forward arrow + 2
END IF
NEXT
```

This should allow the WAO II to follow the right wall, back up when it touches it, turn the other way, and continue. The cliff avoidance routine stops the unit from backing up and turning too far, and turns it back to the right. The forward motion is linked to the right sensor, and moves the robot forward when this sensor is off (no wall is sensed). You may have to adjust the dimensions of the platform you build, and the command sets (length of turns and forward motion especially) for the best effect.

For all of these programs, make sure you enclose your IF–END IF statements properly. The continuous forward motion can be adjusted by increasing the value in the Forward arrow section of the above programs. Note that setting a low forward motion control causes the WAO II to stop and start a lot (less time for forward motion). Adjust this accordingly for your own programs.

Programming WAO II with a PC

You can program the WAO II via a PC with a special interface cable and software available from Mondo-Tronics' Robot Store (www.robot store.com). This is a small cable that connects the parallel port of your PC and the computer interface on the WAO II. The programming language is Microsoft Basic. Use the Microsoft QuickBasic application to run the sample files, and create your own programs.

Programming the WAO II using BASIC is similar to programming it via the keypad. The commands are stored in BASIC files, however, where you can modify them (to tweak settings, for example) without having to hit all those buttons. The interface uses a subroutine to submit the commands to the WAO II one at a time. The computer beeps as each instruction is loaded, to verify that it was received.

Note that the programming interface doesn't control the WAO II directly, and you have to download an entire program into the robot's memory in order to run it. Still, it does make it easier to write more complex programs and test programs that you need to adjust.

The WAO II BASIC interface uses the command values:

```
C = X; A = X
```

where C = is the command prefix and A = is the amount prefix. For example, the command set C = 9, A = 2 tells WAO II to move forward two units, since command number 9 is the BASIC equivalent for forward motion, and amount number 2 is the motion variable for two movement units.

Each program and movement command is assigned a number and has specific amount variables it can accept. See Table 1 for more information.

The number code is a 4-bit data string that passes the command to the microprocessor (for example, 1 = 0001, 2 = 0010). You don't have to program with the string values, since the interface does the interpreting automatically.

A simple square-drawing program converted into BASIC would have a result like the square-drawing program given on page 128). The BASIC equivalent would be the following:

```
C = 9: A = 2
C = 11: A = 1
C = 8: A = 4
C = 11: A =1
C = 9: A = 2
C = 11: A = 1
C = 8: A = 4
C = 11: A =1
C = 9: A = 2
```

```
C = 11: A = 1
C = 8: A = 4
C = 11: A =1
C = 9: A = 2
C = 11: A = 1
C = 8: A = 4
C = 11: A =1
```

A sample program for the WAO II shows that you can also add comments to your command lines, to help you keep track of things (all statements preceded by an apostrophe (') are ignored). Also note the addition of programming that will download the command sets to your WAO II. You shouldn't have to modify those lines. There's also a list of commands you can cut and paste into your own programs at the end.

Number Code	Command	Amount of Variables
1	FOR	1–9 (infinity)
2	NEXT	1 only
3	IF	1–SLON (sensor left on). 2–SRON (sensor right on). 3–SLOFF (sensor left off). 4–SROFF (sensor right off).
4	END IF	1 only
5	Left turn	1–9
6	Right turn	1–9
7	Left pivot	1–9
8	Right pivot	1–9
9	Forward arrow	1–9
10	Back arrow	1–9
11	Stop	1 only
12	Buzz	1–9
0	End Download	none

Table 1. WAO II Commands.

Here's a sample right wall-following program from Mondo-Tronics'.

RWALL2.BAS

```
'WAO II Basic Program - RG 9609.28
'Modified by Richard Raucci 1999
\
'RWALL2.BAS
\
'Right-hand wall follower
\
'Keeps the wall on right side and turns as needed.
\
'Remove left sensor wire.
'Mount only right sensor wire, position at 45
degrees up from horizontal.
\

'SETUP VARIABLES AND PARALLEL PORT
        ON ERR GOTO Done: DEF SEG = 0
        Plport = PEEK(1032) + (256 * PEEK(1033))
        CLS : OUT Plport, 0: OUT Plport + 2, 0

Start:  PRINT "1) Connect cable from computer
parallel port to WAO II"
        PRINT "2) Press Reset on WAO II, red LED
should light"
        INPUT "3) Press Enter on computer to start
download, or Q to quit:", A$
        IF A$ = "q" OR A$ = "Q" THEN END

'WAO II DATA SECTION
\
'Copy and paste the lines at end of this file
'into this section and modify the A values
'for your own programs
'24 steps maximum

C = 1: A = 9: GOSUB Sendit       'Command FOR, Amount =
1 to 9 (9=Infinite loop)

C = 3: A = 4: GOSUB Sendit       'Command IF, Amount:
1=SLON 2=SRON 3=SLOFF 4=SROFF
C = 6: A = 1: GOSUB Sendit       'Command RIGHT TURN,
Amount = 1 to 9
 C = 3: A = 2: GOSUB Sendit        'Command IF, Amount:
1=SLON 2=SRON 3=SLOFF 4=SROFF
C = 7: A = 2: GOSUB Sendit       'Command LEFT ROTATE,
```

```
Amount = 1 to 9
C = 4: A = 1: GOSUB Sendit      'Command ENDIF, Amount
= 1 only
C = 4: A = 1: GOSUB Sendit      'Command ENDIF, Amount
= 1 only

C = 3: A = 4: GOSUB Sendit      'Command IF, Amount:
1=SLON 2=SRON 3=SLOFF 4=SROFF
C = 9: A = 3: GOSUB Sendit      'Command FORWARD,
Amount = 1 to 9
C = 4: A = 1: GOSUB Sendit      'Command ENDIF, Amount
= 1 only

C = 3: A = 2: GOSUB Sendit      'Command IF, Amount:
1=SLON 2=SRON 3=SLOFF 4=SROFF
C = 10: A = 1: GOSUB Sendit     'Command BACK, Amount
= 1 to 9
C = 7: A = 2: GOSUB Sendit      'Command LEFT ROTATE,
Amount = 1 to 9
C = 4: A = 1: GOSUB Sendit      'Command ENDIF, Amount
= 1 only

C = 2: A = 1: GOSUB Sendit      'Command NEXT, Amount
= 1 only

C = 0: GOSUB Sendit             'Always last line of
download

'DOWNLOAD COMPLETED

Again:   IF DelayCnt > 120 THEN PRINT "Error - check
the connections!"
         PRINT ""
         IF DelayCnt < 120 THEN PRINT "Unplug the
cable and press RUN"
         PRINT ""
         OUT Plport, 0: OUT Plport + 2, 0: GOTO Start
'Reset Port

Done: END

'SUBROUTINES

Sendit: 'Download Command and Amount with proper
handshaking
         DlyCnt = 0                 'Clear Delay Counter

CheckDelay1:                        'Download Command
```

```
(C) value
        IF DlyCnt > 120 THEN GOTO Again
        StartT = TIMER          'Set Start Time
CLoop1: EndT = TIMER: IF EndT - StartT < .01 GOTO
CLoop1   'Wait 1/100 sec
        Busy = INT(INP(Plport + 1) / 128)
'Check Busy Line
        IF Busy = 0 THEN DlyCnt = DlyCnt + 1: GOTO
CheckDelay1:
                                'Loop until Busy
Line is HI
        OUT Plport, C           'Put Command Data on
parallel port
        OUT Plport + 2, 1       'Set Strobe Line HI
        DlyCnt = 0              'Clear Delay Counter
CLoop2: IF DlyCnt > 120 THEN GOTO Again
        Busy = INT(INP(Plport + 1) / 128)
'Check Busy Line
        IF Busy = 1 THEN DlyCnt = DlyCnt + 1: GOTO
CLoop2
                                'Loop until Busy
Line is LO
        OUT Plport + 2, 0       'Set Strobe Line LO
        IF C = 0 THEN RETURN        'if last
line of download
        DlyCnt = 0             'Clear Delay Counter

Checkdelay2:                    'Download Amount (A)
value
        IF DlyCnt > 120 THEN GOTO Again
        StartT = TIMER         'Set Start Time
ALoop1: EndT = TIMER: IF EndT - StartT < .01 GOTO
ALoop1   'Wait 1/100 sec
        Busy = INT(INP(Plport + 1) / 128)
'Check Busy Line
        IF Busy = 0 THEN DlyCnt = DlyCnt + 1: GOTO
Checkdelay2:
                                'Loop until Busy
Line is HI
        OUT Plport, A           'Put Amount Data on
parallel port
        OUT Plport + 2, 1       'Set Strobe Line HI
        DlyCnt = 0             'Clear Delay Counter
ALoop2: IF DlyCnt > 120 THEN GOTO Again
        Busy = INT(INP(Plport + 1) / 128)
'Check Busy Line
        IF Busy = 1 THEN DlyCnt = DlyCnt + 1: GOTO
```

```
ALoop2
                                     'Loop until Busy
Line is LO
        OUT Plport + 2, 0          'Set Strobe Line LO

RETURN   'end of Sendit subroutine

'LIST OF COMMANDS
'
'Copy and paste the lines below into the
'WAO II DATA section above and modify the A values
'for your own programs

'C = 1: A = 1: GOSUB Sendit       'Command FOR, Amount
= 1 to 9 (9=Infinite loop)
'C = 2: A = 1: GOSUB Sendit       'Command NEXT, Amount
= 1 only
'C = 3: A = 1: GOSUB Sendit       'Command IF, Amount:
1=SLON 2=SRON 3=SLOFF 4=SROFF
'C = 4: A = 1: GOSUB Sendit       'Command ENDIF,
Amount = 1 only
'C = 5: A = 1: GOSUB Sendit       'Command LEFT TURN,
Amount = 1 to 9
'C = 6: A = 1: GOSUB Sendit       'Command RIGHT TURN,
Amount = 1 to 9
'C = 7: A = 1: GOSUB Sendit       'Command LEFT ROTATE,
Amount = 1 to 9
'C = 8: A = 1: GOSUB Sendit       'Command RIGHT
ROTATE, Amount = 1 to 9
'C = 9: A = 1: GOSUB Sendit       'Command FORWARD,
Amount = 1 to 9
'C = 10: A = 1: GOSUB Sendit      'Command BACK, Amount
= 1 to 9
'C = 11: A = 1: GOSUB Sendit      'Command STOP, Amount
= 1 only
'C = 12: A = 1: GOSUB Sendit      'Command BUZZ, Amount
= 1 to 9
'C = 0: GOSUB Sendit              'Must always be the
last line of download
```

Use the BASIC programming codes shown to enter in your own programs. You can start with the ones we described above for keypad programming. It's a lot easier to write a maze-solving program in BASIC, for example, and adjust it by changing a simple variable and downloading the new program.

Special Projects

The conditional sensors and the piezoelectric buzzer could be used to make more complex routines for the WAO II. One example is a program that would cause the buzzer to sound when both sensors are on (indicating a potential jam in a right-angle corner, perhaps, as the sensors turn WAO II back and forth between the other walls). The buzzer could also be used in conjunction with a sound-activated switch (not included with the WAO II) to activate a door or lower a bridge.

For the door example, have WAO II follow a right wall with both sensors in place. Construct a corridor with a sound-activated door at one end. The buzzer would sound when the WAO II unit reached the door and hit it with both sensors. The robot would back up, stop for a bit, and give off a peep. The sound switch would open the door, and the WAO II would proceed.

The drawbridge example would use the cliff-avoidance program, and modify it to make a sound when it reached the edge of a raised bridge. The sound would activate the bridge, and the robot would wait for it to come down before proceeding.

Both examples show how it could be possible for you to develop behaviors for a robot like WAO II. In these examples, not only is WAO II sensing the environment and moving accordingly, but it can also issue its own commands (with the proper setup).

Documentation

OWI does its usual fine job with the WAO II manual. The kit-building instructions are clear and profusely illustrated. The elaborate electrical circuit diagrams and mechanical action drawings aid understanding about how robotic systems interoperate. The only thing lacking is more information on how to program the WAO II (for example, where is the information on how to draw short phrases?), and better ideas on what to do with it once you have it up and running.

You do get some programming information in the pamphlet that comes with the WAO II PC interface cable (available from Mondo-Tronics'). It also illustrates how to connect the cable to the WAO II and includes a cut-out label to help keep the connector right side up. The programs in BASIC on the accompanying disk should also be read, since they go further in describing interesting things you can do with the WAO II.

Final Thoughts

The WAO II is an excellent multipurpose robot for going beyond the basics. The robot's main drawbacks are its small size and relatively low memory, although the WAO II is capable of sophisticated behavior.

It is interesting to work with a robot that can use conditional programming to make decision-based actions. The merging of sensor-based robotics with programmed action combines the two elements that separate a real robot from a mechanical toy.

Also be on the lookout for a newer version of this robot, the WAO II G, which lets you explore fuzzy logic (a more natural way to create robot programs). See the OWI web site for more information.

Contact Information

Navius
$70

WAO II
$75 (requires soldering)
$99 (no soldering)

WAO II PC Interface
$29

OWI Elekit
1160 Mahalo Place
Compton, CA 90220-5443
(310) 638-8347
www.owirobot.com

Also available from:

Mondo-Tronics' Robot Store
4286 Redwood Hwy, PMB-N #226
San Rafael, CA 94903
(415) 491-4600
www.robotstore.com or www.Mondo.com

Robix

T he Robix™ RCS-6, from Advanced Design, Inc., is an innovative kit that shows how servos work in robot systems. Servos are motors that allow more precise control in motion. For that purpose, most of the models in this reconfigurable kit are centered around a multifunction robot arm. There are some limited walker models also described, and some unique types (like a robot snake). The smooth, direct motion of the servos has to be seen to be believed.

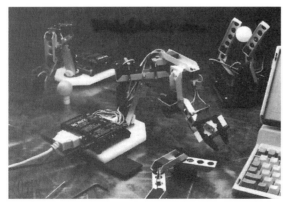

The Robix™ RCS-6 robot kit models (Golfer, Chemist/ Coffee Maker, Snake, and Fingers).

Features

The kit consists of a sturdy platform, custom aluminum building sections, the servos, and a controller box. The controller has a CPU that acts as an interface between a PC and the robots you create. It can learn and save repeatable macro files that act as the robot's programs. The kit is very complete – a storage box, safety glasses, and an instructional video for all models are included. The controller is preassembled, so there's no soldering involved.

The CPU box must be attached to a PC (there's no autonomous battery action in this version), so your robots will be limited to a short range. This is less important with the sophisticated arm models, since they operate from a fixed position.

No sensors are included in the kit, but there are instructions for adding them, a fairly easy task, and you can buy them as an option – both touch and light kinds.

The software works well, and has a graphical user interface (GUI), used for teaching and directly controlling the robots. Programming via C and Quick Basic Version 4.5 is more challenging, and you should refer to the manual for more information.

None of the models have macro files predefined on disk, so you'll have to write your own sample programs into a computer file. Most of the models have sample programs printed in the instruction manual that you can type in and modify. You'll have to write your own programs to match the robots that you create based on the examples.

Program Mode

In Program Mode, you'll use the software to write a program that controls the servos. The Robix™ RCS-6 uses a saved macro structure that compiles routines into the device driver (the software program that drives the Robix robot model that you build). No separate control files are saved, but you can pick a macro by name once it is loaded into the driver. You can run macros either from the command line in the console program or from a menu.

Teach Mode

Macros are saved line-by-line using the Teach panel. After you move one of the servo motors using the keyboard, press Alt-A to place the

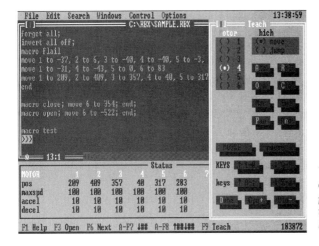

The DOS GUI console program for the Robix™ RCS-6 kit.

instructions for that motor into the console window. There are options for fine and coarse motor control and absolute or relative moves. Generally speaking, an absolute move will start from the same place each time, and a relative move will move from the last stopping point.

You can also run macros multiple times by adding a number value. For example,

```
test 3
```

will run the test macro three times. Use the 0 value to run the macro indefinitely.

To compile the macro into the device driver, first save a string of servo commands using the Teach window. Then, put the cursor on

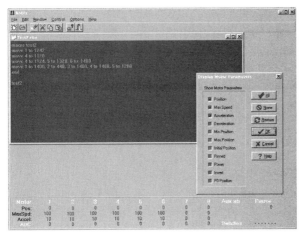

The new Windows 95 Robix console program.

the top line. Enter the macro command and a name on the top line, and the end command on the last line. Then go back to the top line, and use the F8 key to step through the commands and put them into the driver.

The example below is used with the Chemist robot arm model. The macro name is on the first line. The next lines use the move commands to move the servos by number. Note that multiple servos can be moved on one line.

```
macro test2
move 1 to 1242
move 4 to 1120
move 4 to 1124, 5 to 1320, 6 to -1400
move 1 to 1400, 2 to 440, 3 to 1400, 4 to 1400, 5 to
1280
end
```

This short program moves the arm around, lowers the arm sections, turns and opens the mechanical claw, and lifts the arm. It was generated by driving the robot via remote control from the keyboard.

It's essential that you take the time to understand which keys run which motors, and where they are located. You should make several practice runs before you compile and save your macros.

Here's a sample script for the Golfer robot, provided by the manufacturer:

```
invert all off;      # set invert to simplify
scripting
invert 3 on  # if using Futaba servos
invert 1,2 on# if using Hitec servos
move 4 to 711# choose the club head angle
forget all # clear out all macros
macro swing
accdec 1,2,3 10; maxspd 1,2,3 30;
move 1 to minpos; move 2,3 to minpos;
pause 25 # pause to let operator load a ball and get
clear.
accel 1,2,3 500; decel 1,2,3 20; maxspd 1,2,3 200; #
start hard, stop easy
move 1,2,3 to maxpos
end ;
swing 0 # swing indefinitely
```

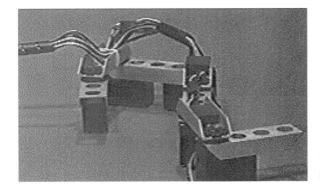

The Snake model.

Example Models

Snake

The Snake uses the servos to create a six-jointed robot capable of sinusoidal movement. It's a neat example of how you can use robotics to explore different types of locomotion. The Snake instructions include a sample program.

Chemist

The Chemist model is a good demonstration of how to use a robot for a practical task. It looks like a strong performer in the accompanying video. You'll have to come up with the program yourself, to match your own layout.

The Chemist model.

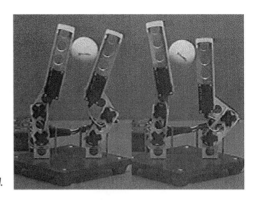

The Fingers model.

Fingers

This example uses the servos in a novel way, to replicate two triple-jointed fingers placed side-by-side as a type of robot hand. Built this way, the robot is capable of fine motor control and allows you to explore ways a robot hand might differ from the human model. For example, these fingers can bend backward at the joint. The Fingers instructions include a good program example.

Golfer

The Golfer is a four-axis robot arm that features rotation around the base, shoulder and elbow control, and an adjustable club angle. It's a good model, and can send a ball about 16 inches. There's a sample program in the instructions.

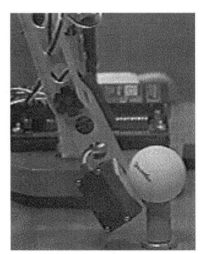

The Golfer model.

The Coffee Maker model.

Coffee Maker

This model does a very nice demonstration of robot functionality – it actually completes a serious multi-step task! It's encouraging to see a robot that can do a four-stage task. It starts with a cup of hot water, then scoops instant coffee from a fixed jar into the cup, then moves to the sugar jar, scoops some up, and puts that in. Next, it ladles milk from a jug into the cup, and for the final step, it stirs it up. Theoretically, all you have to do is to keep the setup where the robot expects it to be, and just place a cup of hot water at the same spot. The robot will do the rest.

Take a good look at the video for this model, for there's no sample program (although there are programming tips). It would be better for the vendor to assume that users will build the model exactly as shown in the diagram, so a sample program would work just fine. They should

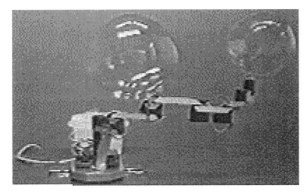

The Bubbler model.

The Thrower model.

also provide information on how to make the stands for the coffee and
the sugar, and tell how far apart to place everything, but you can modify
the model to fit your own plans.

Bubbler

Bubbler is a fun demo robot arm that makes large-sized bubbles by dip-
ping a wand into bubble soap and whisking it around its base. It's an
interesting alternative function for a robot. Use the programming notes
to get this model up and running.

Thrower

The Thrower arm shows how fast a robot can operate. It pitches ping-
pong balls into a cup with good accuracy. This one is great for showing
off your robot's capabilities.

Dancer

Despite the low-quality video, this model shows how the multiple servos
in the Robix™ RCS-6 could be used to animate a puppet with a high
degree of control.

The Dancer model.

The Strider model.

Strider

The Strider is a two-legged walker that is simple to build. It's of limited use, but it does show how a walking robot could work.

3-Legs

The 3-Legs model is a bit clunky, but it's the only example that actually carries the computer interface, making it a bit closer to an autonomous robot. It does have to stay attached to the PC and the AC power supply to receive commands and power. Use this one to see how servos can be used in a walker platform.

Draw-Bot

This arm model uses servo control to draw with a pen. You can use it to draw repeated shapes with a macro program, or more complex drawings

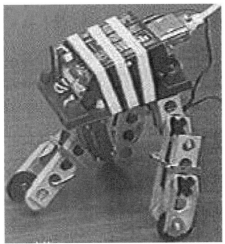

The 3-Legs model.

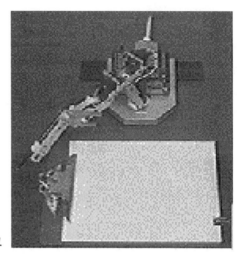

The Draw-Bot model.

using the Teach method of direct control, which can then be repeated on its own indefinitely.

Further Suggestions

You can add sensors to the above models to increase their functions. The sensors are available from Advanced Design, and plug into the controller. The general idea is to use the sensors with the example models – for example, you could use the sensors to change the motions of the arm models (the arm would move toward or away from a light source).

Try building a better walker. Note that you'll still be limited to the line connecting the controller to the PC, but you can experiment with alternate leg movements and the number of limbs.

See if you can make a multifunction tabletop robot – a Chemist and a Coffee Maker in the same model without rebuilding, and use different macro programs (and arm attachments) to change the functions.

Build whatever you want – the kit is a full robot construction set, and you can go beyond the example models once you understand the principles involved.

Final Thoughts

This is an interesting kit. You can learn a lot about robotics from it, and the servo motors and the custom aluminum parts work together well.

The programming interface is unique, and seems to be well placed between the text editor-based downloadable C files for robots like the Rug Warrior Pro™ (see Chapter 12) and the easier graphical programming software for robot kits like FischerTechnik Mobile Robots (see Chapter 11) and LEGO MindStorms (see Chapter 3). The Teach method of programming the Robix robots by example is a good idea, something other robot manufacturers should think about including with the software for their robots.

Contact Information

Robix™ RCS-6
$550
6052 N. Oracle Rd.
Tucson, AZ 85704
(520) 544-2390
www.robix.com

FischerTechnik Mishike Robots Kit - # 30400

T he Mobile Robots kit from FischerTechnik is the latest in a series of kits built around their Intelligent Interface, a battery-powered CPU with 32 kilobytes of RAM that interfaces with motors and sensors to create complex mechanisms. The kit also features 270 FischerTechnik parts that can be used to construct five robots, programmed using Windows-based software. The interface works in conjunction with the software to run programs that you download.

The Mobile Robots kit.

The FischerTechnik Intelligent Interface.

The kit has a wide range of interconnectable FischerTechnik parts, including lots of connectors, two motors, a set of wheels, and two types of sensors (light and touch).

The Intelligent Interface can control up to four motors and eight digital sensors (sixteen with an optional adapter), as well as two variable analog ports. It connects to a PC via a serial cable (included) and stores a program in memory as long as the power is on.

This is a great kit because it outlines several principles of robotics in five main models, and it's also reconfigurable, so you can build your own robots.

Models You Can Build

MR1 – Basic

The first model is a basic two-wheeled robot. You can program its movements via the Lucky Logic robot control software. This will give you a

The Mobile Robot kit's main models based on the Basic Model.

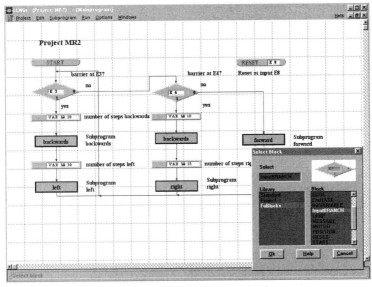

The Lucky Logic (LLWin) robot control software.

good idea of how to program general forward, backward, and turn movements for your robot. This is also the base platform for the other models that you can build.

MR 1 – Basic robot.

MR2 – Obstacle Avoidance

This model is an obstacle avoidance robot. You build it by connecting a bumper to the front of the robot, with two sensors on either side behind the bumper. The example program sets the MR2 to wandering about a room. When the left side of the bumper hits an obstacle, the robot backs

MR2 – Obstacle avoidance robot.

up and turns right. When the left side is touched, the opposite action takes place. In this way, the robot will cruise about, correcting its course as necessary. It will eventually find the exit in a room with one doorway.

This is a good model to modify for maze-following. You can change the robot's program to follow a left or right wall by disabling the opposite sensor and tweaking the movement commands.

MR3 – Edge Detection

The MR3 model uses the microswitch touch sensors to create a robot that senses ledges. As long as the two feeler arms make contact with the floor, the robot will continue its movement. If one drops over the side, the switch is engaged, and the robot will back up and make a course correction away from that side.

This action shows how a robot can be constructed to be aware of abrupt changes in its environment. A general purpose for this type of sensor setup would be to make sure that a robot doesn't exceed a bound-

MR3 – Edge detection robot.

MR4 – Light-following robot.

ary (i.e., fall down a flight of steps, or fall off a tabletop). You can also easily reconfigure these sensors to point upward, and the same program will keep the MR3 robot under a low table (just make sure the sensors make contact with the underside of the table).

MR4 – Light Detection and Following

By adding a pair of light sensors to the basic model, you can make a light-following robot. The MR4 uses the two photosensors to scan for light, then locks onto a bright light source, and heads toward it.

A light-detecting robot could be used to find a lit doorway, or as a part of a firefighting robot (to move toward the source of the flame). You can also modify the example program to change the robot's behavior – for example, you could easily program it to avoid a light source, turning and heading in a different direction.

MR5 – Line Following

The MR5 follows a dark line, turning with it, and staying on the exact path that you set down for it. It uses two light sensors mounted left and right at the bottom of the robot, along with a light source. The sensors detect the line on either side, making course corrections as necessary.

This robot shows how you can use sensors for alternative programming. The basic program tells the robot to move left or right if a dark area is detected by the light sensor on the opposite side. The path program is laid down using black electrical tape on a white background (or you can use a thick black magic marker). Possible uses include a robot delivery route or a reconfigurable navigation system (make sheets of paper with straight lines and curves on them, and line them up in different ways to move the robot about).

MR5 – Line following robot.

Programming

The programming interface for the Mobile Robots kit is the LLWin software, a Windows-based visual application that works with the Intelligent Interface (included with the kit). Each program is made up of function blocks and looks a bit like a circuit diagram. You put the program together by assembling the blocks and linking them together. In this way, you can set the movements for the robots and also the sensor conditions that will affect their behavior.

For example, the program for the MR2 – Obstacle Avoidance robot was shown at the beginning of the chapter (see pages 149-150). The Start block leads to the two Variable blocks, which are tied to the two sensors on the bumper by name (E3 and E4) and act as decision-makers. If the sensors are both disengaged (a "0" state), the program branches off to the forward motion subprogram, and the robot travels forward. The variable blocks branch out separately when one or both of them are contacted (a "1" state) to run the relevant left or right subprogram (which backs the robot up and turns it on a different course, to avoid the obstacle).

You lay out the function blocks using the pull-down menus, connect them using the mouse, and edit the programs using pop-up dialog boxes. You can also label your program using a built-in text editor. In the example, the labels have already been added, to make things clearer.

After you have tailored the program to your robot, you download it using the Run/Download command, and your robot is ready to go. The Mobile Robots kit comes with sample programs for the models described, so you can get started, and you can edit them to see how the changes will affect their behavior. For example, if you notice that your MR2 robot is getting stuck in corners, you may want to change the number of steps left, right, and backward to get the robot to perform better.

Information on programming the Intelligent Interface via Visual Basic is at a FischerTechnik related web site, http://ourworld.compuserve. com/homepages/UlrichMueller/fishface.htm (in German; translate it by going to http://babelfish.altavista.digital.com/).

Further Ideas

The models in the Mobile Robot kit are fine examples of general robot principles. It would also be interesting to see how the interaction between sensors could be addressed by this kit. As it stands, the four robot models that use sensors fall into two groups: The MR2 and MR3 robots use the touch sensors, and the MR4 and MR5 robots use the light sensors.

You could conceivably construct a robot with both sensors – for example, a version of the MR2 — Obstacle Avoidance robot that would also have a light-following behavior. This robot would wander about a room, bumping off obstacles, unless it detected a strong light source, which it would home in on.

Try adding a subprogram to a copy of the MR2 example program, just after the Start block, so that a light-sensor subprogram would interrupt the obstacle detection program if it sensed a light, but let it run if no light was present. The robot would wander around a room, unless it "saw" a lit doorway, whereupon it would head straight for the exit. In this way, the robot would appear to be capable of figuring out the fastest route out of the room on its own.

Use the MR4 program for the basic light-sensing routines, and refer to the documentation on the CD-ROM for more information. You may also want to try this sensor interaction with other robots of your own design.

Remember that the FischerTechnik kit is fully modular, and you can use the pieces in the kit to make any kind of robot you can think up (a walking robot, for example). The analog inputs are especially interesting, since you can add heat and brightness sensors to expand your robot's capabilities. See the contact information below for ordering additional kits and special parts (like additional sensors).

Final Thoughts

FischerTechnik gets high marks for providing a well-thought-out robot kit, with models that perform well and illustrate strong robotic principles. The LLWin software is one of the best robot control applications in its class. It has a high level of functionality and is fairly easy to use. This is a great kit, hands down.

Contact Information

FischerTechnik Mobile Robots (#30400)
$249

fischerwerke
Artur Fischer GmbH & Co. KG
D-72178 Waldachtal
0 74 43/12-4 369
www.fischerwerke.de

Distributed in the United States by:

Model A Technology
2420 Van Layden Way
Modesto, CA 95351
e-mail: modelatech@telis.org
(209) 575-3445

Related Web Sites:

http://www.knobloch-gmbh.de/fischer/fi-30400.htm (in German)

Rug Warrior Pro™

Chapter 12

T he Rug Warrior Pro™ kit from A K Peters, Ltd. (Natick, MA) evolved directly from the book *Mobile Robots: Inspiration to Implementation*, first published by A K Peters in 1993, which is now available as a paperback in a second edition. It's a complete robot kit that gives you a good background in robotics and benefits from its association with an excellent book that goes into the theory behind it.

The Rug Warrior Pro™ kit from A K Peters.

The heart of the robot is a Motorola MC68HC11 processor, as used in other robots in this section. It uses 32 kilobytes of battery-backed up RAM (random access memory) to store programs downloadable through a serial port PC interface (software and interface cables are included).

This central processing unit (CPU) is connected to a series of sensors and drives two motors through a shaft encoder subsystem. The sensors include dual infrared (IR) detectors, light sensors, three microswitch bump switches, and a sound sensor (a standard microphone). The CPU board also has a small programmable piezoelectric buzzer. Status is reported through a two-line liquid crystal display screen, which is helpful for testing and calibration.

Rug Warrior Pro's body is centered around a circular skirt that houses the three bump sensors, spaced evenly around the perimeter. This skirt gives the Rug Warrior Pro™ the ability to sense by touch in roughly 360 degrees. The robot is 7 inches in diameter × 4.5 inches tall.

The CPU circuit board is located at the top inside the circle, along with the main sensor arrays, and the power/download/reset switches. Underneath this are the two motors and the drive wheels, balanced by a caster, as well as the battery compartment (for six C cells).

The infrared sensors allow Rug Warrior Pro™ to sense its proximity to an object without actually touching it. This is like using a visual signal to navigate through an area. The light sensors can be used to program the Rug Warrior Pro™ to move toward or away from a light source.

The Rug Warrior Pro™ also has expansion capabilities. You can add a sonar board (available separately from A K Peters), or your own sensors. The kit is recommended for experienced electronics builders.

The Rug Warrior Pro™ with optional carry strap attached to it.

The robotic principles Rug Warrior Pro™ illustrates include a wide range of behaviors. The multiple sensor arrays allow you to combine sensor programming to produce interesting effects. The robot's programming uses sensor input to modify its behavior. For example, the Wimp program uses the bump sensors to tell the robot to turn away from the side that gets bumped and travel away from the object that hit it. The behavior is an illustration of the principle of avoidance.

More complex programming can use one sensor input to override another, for more complex behavior. For example, the robot could be programmed to travel toward the nearest light source, unless the noise level in that area is too high. In that case, Rug Warrior Pro™ would override its light-following behavior in order to avoid the harsh sound.

In this example, the robot would be programmed to locate the nearest bright light source by using the onboard photodetectors, which would tell it to turn toward the side that was receiving the most light while driving forward. The microphone sensor would be used in a subprogram to tell the robot to override the light-following behavior when the ambient noise level reached a certain point.

A practical application of this type of complex behavior is to imitate the way humans and animals use sensory input to redirect their behavior. For example, a person would direct him- or herself to a light source in a room in order to be able to read a book, but would steer away from the noisiest part of the room, even if it had the most light, to effect a compromise solution.

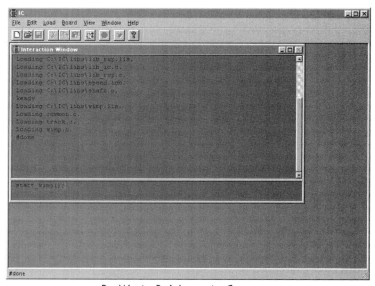

Rug Warrior Pro's Interactive C program.

The Wimp.C program in WordPad.

Programming Essentials

The Rug Warrior Pro™ uses Interactive C for programming. The kit includes Windows 95 (and alternatively, Macintosh or Unix) software that allows you to download programs to the robot. The programs can be edited in any text editor, and are stored and executed by the CPU.

To edit a program, open the relevant file in a program like WordPad. In this example, we'll load the Wimp.C file from Rug Warrior Pro's sample programs library directory.

```
/* Wimp.c — move away from any touch
*/

float oct = 440.0;

float wfreq(int step)
{ if (step == 0)
    return oct;
  else
    return (oct * (2.0 ^ ((float) step / 12.0)));
```

```
}

int wimp_active = 0;        /* Don't start process
when already running */

void start_wimp_tune()
{ if (! wimp_active)
    start_process(wimp_tune());
}

void wimp_tune()
{ wimp_active = 1;
  oct = 1760.0;
  tone(wfreq(5),0.1);
  tone(wfreq(4),0.1);
  tone(wfreq(5),0.1);
  tone(wfreq(4),0.1);
  wimp_active = 0;
}

/* Move away from any bump — play an annoyed tune */

void wimp()
{ int bmpr;
  printf("Wimp\n");
  while(1)
    { bmpr = bumper();
      if (bmpr != 0)        /* Somehow, we were
bumped */
      { start_wimp_tune();       /* Play I'm-an-
annoyed-wimp tune */
        if (bmpr == 0b110)     /* Back and Left
bump */
          track(0,-80,5);      /* Spin right ~60
deg */
        else if (bmpr == 0b010)  /* Left Bump */
          track(0,-80,10);       /* Spin Right 120
deg */
        else if (bmpr == 0b011)  /* Left and Right
=> Front */
          { track(-80,0,4);      /* Backup then turn
away */
            track(0,80,17); }
        else if (bmpr == 0b001)  /* Right Bump */
          track(0,80,10);        /* Spin Left 120
deg */
        else if (bmpr == 0b101)  /* Right and Back
```

```
bump */
                track(0,80,5);          /* Spin Left 60 deg
*/
                track(80,0,8);          /* Go forward a bit
*/
                sleep(0.75);            /* Let bumper damp
out */
        }
     }
}

void start_wimp()
{ tone(1000.0,.1);
  wimp_tune();
  sleep(1.0);
  init_velocity();                 /* Needed by track */
  start_process(wimp());
}
```

Note the commands relevant to Rug Warrior Pro's mission for the program example above. The bumper sensors are denoted by the (bmpr) tag and trigger the shaft-encoded motors by using the (track) tag. This tag controls the motion of the wheels, using a disk with black marks to track and control the amount of travel. Other programming commands include the code necessary to generate the annoyed tune Rug Warrior Pro™ produces when disturbed.

To edit this program, modify the values associated with the track tag. For example, if you want to make an Attack program, change the track values to tell Rug Warrior Pro™ to spin around and run toward the object that has hit it. Save the file under another name (like Attack.C). You'll also have to create a new .LIS file.

The .LIS file is a list of programs that will download into Rug Warrior Pro™. These programs include standard library functions that need to be loaded into Rug Warrior Pro™ to set it up so that the program will run. The Wimp.C program has the Wimp.LIS file associated with it. It looks like this:

```
common.c
track.c
wimp.c
```

In a text editor, open the Wimp.LIS file, and edit it to include the new Wimp.C program. It should look like this:

```
common.c
track.c
attack.c
```

Save the file as Attack.LIS, and you're ready to download the new program into Rug Warrior Pro™ and try it out.

See the Library Functions section in the Rug Warrior™ Assembly Guide (Appendix A) for a complete list of programmable functions.

For an overview of arbitration between behaviors, see Chapter Nine in *Mobile Robots: Inspiration to Implementation*. Also check out the example programs below, and see what modifications you can come up with.

Sample Programs

Bugle

This program uses Rug Warrior Pro™ to play tones when its bump sensors are touched. The frequency of the buzzer sound changes depending on which sensor is touched (and whether the sensors are touched singly or in pairs), and produces discrete notes.

Theremin

Theremin is a cool program that emulates a Theremin, a musical instrument that is played without touching it. A Theremin uses the change in an electrical field to produce sounds as the operator approaches and moves away from the unit. For Rug Warrior Pro™, the Theremin program uses the photocell sensors to control output via the Rug Warrior Pro's speaker. By moving your hands over the light sensors, the amount of light detected is changed, and this changes the pitch and tone repetition rates. You can use this method to play music with your hands.

Yo-Yo

Yo-yo is a simple run-and-return program that moves Rug Warrior Pro™ forward, turns it around, and brings it back to the same spot when the back touch sensor is pressed. You can use the base code from this program to add a return subprogram to larger programs, and it's also useful to make sure that your Rug Warrior Pro™ is calibrated properly.

Wimp

The Wimp program, discussed above, is a beginning example of robotic behavior. The idea behind the program is to have Rug Warrior Pro™ shy away from anything that touches its bump skirt. Pressing the back switch will cause the robot to run forward. Pressing either of the side switches will cause the robot to turn away from an object on that side and then move forward, away from the object.

Follow

This program uses the IR sensors to locate and follow the nearest object. You can use your hands to move the robot left or right, by placing them over the sensors. When Rug Warrior Pro™ touches an object that it is tracking, the bump switch tells it to turn off its motors.

You can use two Rug Warrior Pros™ to run the Wimp and Follow programs simultaneously. The robot running the Follow program will track the Wimp robot, and when it catches up to it, the Wimp robot will run away, causing the Follow robot to track it again, repeating the cycle.

Moth

The Moth program uses the photoelectric sensors to follow the brightest light source. You can move Rug Warrior Pro™ around using a bright flashlight, and the robot will track the light. The bump sensors will stop the motors when Rug Warrior Pro™ strikes an object.

Echo

The Echo program uses the onboard microphone to count sounds, then repeats the number it hears with audio and motion feedback. You can clap three times in succession, and Rug Warrior Pro™ will beep and move forward and backward the same number of times. This is useful to see how a robot can demonstrate giving feedback from input it receives. In this example, Rug Warrior Pro™ is giving a simple indication that it has correctly heard the three sounds. Modify this program to see how specific input can be translated into measured action.

Sonic Commander

The Sonic Commander program is a modified version of the Echo program. In this case, Rug Warrior Pro™ is counting the number of sounds it hears and assigning an action to them. If it hears two sounds, it spins; hearing three sounds causes it to move forward; and hearing one sound causes Rug Warrior Pro™ to stop. The example given in the manual suggests you can emulate simple speech recognition with this program. For example, two sounds ("Please spin!"), three sounds ("Go forward now!"), and one sound ("Stop!") should produce the appropriate action. You may have to modify the program to get the proper results, and you can also assign different actions to the sound input.

Behave

The Behave program loads all of the preceding programs into Rug Warrior Pro™ and allows you to access them by using a menu that shows up in the LCD status window. You may want to run this program first, to investigate the range of programming Rug Warrior Pro™ has to offer.

Special Projects

The Rug Warrior Pro™ Assembly Guide includes suggestions for advanced Rug Warrior Pro™ programs, based upon the example programs provided with the Interactive C software. Note that there is no sample code for these programs, so you'll have to write them yourself. The concepts are straightforward, however, and the instructions are clear.

Single Rug Warrior Pro™ Projects

These projects work for one Rug Warrior Pro™.

Lewis and Clark

This Rug Warrior Pro™ program uses three simple behaviors to cause the robot to wander about a room, exploring at will. The Cruise program tells the robot to move forward. The Avoid program uses the IR sensor input to tell the robot to detect obstacles and change course, overriding the Cruise behavior. The final element is the Escape program, used to keep Rug Warrior™ from getting stuck. It uses the bump sensor to tell Rug Warrior Pro™ if it is jammed against an object like a wall, then backs it up and turns it, enabling the robot to escape being trapped.

Combined, these subprograms will let your Rug Warrior Pro™ navigate a room independently, avoiding obstacles, and keeping out of corners. It's an impressive demonstration of the robot's capabilities.

Barishnikov

The idea behind this program is to choreograph the sound and motion activities of Rug Warrior Pro™ to create a music and dance routine. In this fashion, the robot could dance to a tune that it plays itself (in a limited fashion).

Mouse

The Mouse program is a classic example of robot wall-following behavior. You can use this idea to program Rug Warrior Pro™ to follow a wall without getting stuck. Program the IR detectors to move forward and sense the walls, turning robot to opposite sides for course corrections. In this way, you can program Rug Warrior Pro™ to find its way out of a maze. Use the bump sensor to turn the motors off if the robot gets stuck, or try an Escape subprogram.

Magellan

This program idea suggests navigation systems for your Rug Warrior Pro™. One method uses dead reckoning, a method of navigation that tracks how far the robot has moved to arrive at a specific spot. For example, you'd start Rug Warrior Pro™ at a set starting point, and measure how many encoder clicks it takes to reach a certain destination. Then you'd write a program that would move the robot to its destination, including the turns it would need to reach that spot.

A drawback to dead reckoning is that you'd need to make sure you started your Rug Warrior Pro™ at the same exact spot each time, as even a little variance would create errors that would multiply over the course. Suggestions for alternate methods of navigation that might work better include wall-following and light-sensing. The idea is that you'd use a sensor to track a path or home in on a target for navigation.

Apollo 13

This idea uses the light sensors in two interesting ways. The first suggestion is to use the two photoelectric sensors to home in on a light. By adjusting the Rug Warrior Pro's course right or left according to the difference in light intensity falling on the two sensors, you can get the robot to move toward the light. This is a standard light-following program, like the Moth example.

The second suggestion is to use the photoelectric sensors to monitor the light intensity. At a lower level, Rug Warrior Pro™ will move toward the light. As the light intensity goes up, the robot will move away from the light. In this way, you can write a program to get Rug Warrior Pro™ to orbit a light source at a fixed distance.

Fire!

This idea is to program the Rug Warrior Pro™ to perform for a Home Robot Firefighting contest, like the one held at Trinity College, Hartford, Connecticut annually, and also at selected regional sites across the U.S. (see the web site www.trincoll.edu/~robot/ for more information).

The general idea is to have your robot move through a series of rooms and home in on a candle. As a practical example, you can program Rug Warrior Pro™ to move through a maze (using a wall-following subprogram) and home in on a light source like a small lamp (using a light-sensing subprogram). This will give you a good sense of what a larger-scale home firefighting robot would need in terms of physical capabilities and programming.

Multiple Rug Warrior Projects

These projects are for more than one Rug Warrior Pro™.

You're It!

This idea is for a game of robot tag. Use two small 9-volt lamps wired to two Rug Warriors™ at Expansion Socket A. The It robot runs a light-seeking program, where it tracks the light on the other Rug Warrior™. The Not-It robot runs a light avoidance program, where it tries to avoid the light on the It robot.

The interesting interaction occurs when the It robot "tags" the Not-It robot by colliding with it. The bump sensor on the It robot signals that robot to change its program to the avoidance behavior, and the bump sensor on the Not-It robot signals that robot to change to the light-seeking behavior. As in the real game, when the Not-It robot is tagged, it becomes It, and the chaser becomes the chased.

Couch Potato

You can use this suggestion to develop a program for inter-robot communication. The general idea is to use the IR emitters on one Rug Warrior Pro™ to signal another. The sensor inputs on the transmitter robot would be connected to its emitter, and the receiving robot would constantly monitor its IR detector for a proper signal. For example, pushing the back bumper on the transmit robot would cause the receiving robot to rotate 90 degrees.

Gentlemen, Start Your Engines!

This programming idea concerns robot racing. The first suggestion is to add a line-following subsystem to your Rug Warrior Pro™ robots. This would allow your robots to follow the same path in a race. Alternate suggestions include a wall-following race, or a light tracking and following contest.

Ready or Not, Here I Come

This is a complex idea for a game of robot hide and seek. The general idea is to use the IR emitter and detectors to allow a Seeker Rug Warrior™ to search for Hider Rug Warriors™. The Seeker would start out by playing twenty beeps from its buzzer. The Hider robots would hear the sound and run away, stopping when they hit an obstacle. Then the Hiders would turn to face the way they had came, and turn on their IR emitters. The Seeker would then spin in place, using its IR detector to sense the Hiders' emitters. If a Hider robot is detected, the Seeker robot homes in on that signal, and bumps into it. The Hider signals the hit by beeping when touched, and the Seeker goes to look for the next Hider.

You'll have to write separate programs for the Seeker and the Hider robots, but you should be able to use the same program for the Hiders (just place them in different spots around the Seeker).

Out of Africa

Robot cooperation is illustrated in this idea. The suggestion is to write programs that will allow your Rug Warrior Pros™ to interact on a task. For example, one robot could monitor its shaft encoders and bump sensors to detect an object (like a box), and tell whether it can move the object itself. If the first robot is applying motor power and the encoder clicks aren't detected, then the box is too heavy. The robot can then either abandon the box and try another one, or call for help, by turning on its IR emitters to signal another Rug Warrior Pro™. You'd have to stack the odds against the first robot to show true robotic cooperation between two Rug Warrior Pros™, by making the obstacle too heavy for one robot to shift alone.

World Cup

Robot soccer is an interesting contest that is gaining in popularity. The general idea is for a set of robots to locate a ball and the opposite goal, and deliver the ball to the goal. Rug Warrior Pros™ can be used to play a form of this game. Some ideas include using an IR beacon for each goal (transmitting different codes), and a small puck with a light inside it as the soccer ball. A whistle starts the game, and the Rug Warrior Pros™ find the ball with their photocell detectors and try to push it into the opposite goal.

The programs for both team of soccer robots would be very similar. The main difference would be the scan codes for the goal beacons, to differentiate the two goals from one another.

Final Thoughts

The Rug Warrior Pro™ is a robust robotics platform. It can be a serious task to build this robot, but the rewards are worth it. The knowledge you gain from the construction of the robot will help you understand how the mechanical elements interact with the electronics.

After you have built the robot, Rug Warrior Pro's range of sensor programmability will let you explore robotics to a wide extent. Good sample programs are included, along with many suggestions on how you can continue to work with it to increase your robotics knowledge.

Contact Information

Rug Warrior Pro™
Brains and Brawn Kit $599
A K Peters
63 South Ave.
Natick, MA 01760-4626
(508) 655-9933
e-mail: service@akpeters.com
www.akpeters.com

Section 3

Advanced
Robot
Platforms

Pioneer
Robots

The Pioneer series from ActivMedia Robots are cool autonomous robots. You'll find that the Pioneer is exceptionally well made. The models come fully assembled. Each features sturdy CNC (computer numeric control) aluminum construction, sonar and bump sensors, and lots of expansion options.

The Pioneer 2.

The Pioneer 2-AT.

The Pioneer 2 CE has a two-wheeled base with 6.5-inch tires and a rear castor. The unit is 6.5 inches tall, and uses an offboard processing system (a host computer) for robotic control (or a piggybacked laptop connected via a serial cable interface). An optional sonar ring provides 8 sonar sensors with 180-degree coverage. It uses a rechargeable battery system for power.

The Pioneer 2 DX features a similar design as the Pioneer 2 CE, including a host computer configuration. It also includes the sonar ring described above as standard. An optional PC104+ interface makes this robot Ethernet-, modem-, and laser rangefinder-ready. (These are more

This picture shows the relative sizes of the earlier Pioneer 1 (bottom left) and the Pioneer AT (middle right).

*Side view of
Pioneer 2 robot.*

advanced options for robotic control and sensing.) It also features a hot-swappable rechargeable battery system (so you don't have to shut the robot down to change batteries), and more powerful motors than the Pioneer 2 CE. It can reach a speed of 2 meters per second, and carry up to 23 kg of payload.

ActivMedia also makes an ATV (all terrain vehicle) model of the Pioneer, with 4-wheel drive and a rugged chassis. The Pioneer 2 AT includes an embedded computer system, providing onboard control for more autonomous robot activity, as well as the PC 104+ interface. It also features a hot-swappable rechargeable battery system, and can carry up to 40 kg of payload (around 90 lbs).

Features

Each model features a Siemens C-166-based microcontroller for storing and executing programs, and includes the Pioneer 2 Operating System (P2OS) to control basic functions. Both CE and DX models interface with a host computer via a serial cable or a radio modem link (either a desktop PC, or a laptop that the Pioneer can carry itself).

Other features include a liquid crystal display for system status and messages and a piezoelectric speaker. Pushbuttons on the top of the case include an on/off switch, motor power control, and access to optional functions.

An interesting part of the Pioneer robot's construction are the integrated sonar sensors (optional on the CE and the AT models). Eight sonar transducers mounted in a 180-degree ring are used to detect objects and features in the Pioneer's environment (like a doorway), as well as to collect range information. The time that the signal takes to bounce back to the sonar unit is directly related to the distance of an object. In this way, the Pioneer can figure out how close it's getting to an object. There's also one extra sonar port on the motherboard for optional use.

The Pioneer uses 500 tick optical encoders on the drive shafts. This allows the robot to track its position to within one percent accuracy. The design of the Pioneer also allows it to turn in place, with minimum space needed.

The Pioneer series comes with interactive programming and operation software developed at SRI (Stanford Research Institute), called Saphira. A custom version of LOGO and alternate programming languages also work, as well as new reactive software called Ayllu.

Options

Gripper

You can buy a robot gripper-manipulator for your Pioneer. It typically functions to pick up and move small objects. The range of motion for the grip paddles is to lower and open, and close and raise. Note that this means that the Gripper only opens in the down position.

The grip paddles are 1 1/2 inches high × 3 1/2 inches deep, and ride 1/4 inch off the floor in the down position. The interior width is 9 inches when fully open. It's controlled by one servo-motor connected to the Pioneer microcontroller board.

The Gripper is made of sturdy aluminum and features foam pads in the jaws. It has infrared and switch sensors in the tips of the paddle

An early Pioneer with a gripper attachment in place.

arms and along the transit of the gripper path (to sense objects and paddle position).

The Gripper is useful for giving your Pioneer the ability to complete tasks. You can use it to make a delivery routine and for other related chores. It's also useful for robot competitions, like Robot Soccer. The Gripper attachment costs $1,495.

Compass

The Digital Compass add-on for the Pioneer gives your robot compass headings. This can significantly expand the robot's abilities to figure out its position and to reach a destination based on directional information. The compass plugs into the Pioneer's serial port via a pass-through plug that allows you to keep other serial devices attached to the robot. The software interface is a plug-in to the Saphira and PSOS (Pioneer Operating System) control programs. The Digital Compass costs $495.

The Digital Compass option.

Vision Systems

PTZ (Pan Tilt Zoom) Controllable Camera System

This is a Sony camera that mounts to the front of the Pioneer and can be moved in three directions to access the zoom function (1X – 12X) by using Pioneer software in your robot's programs. It's not a vision system per se, but it will give your robots surveillance capabilities. It includes a 2.4 gigahertz (gHz) AV Pro transmitter–receiver set to send the video back to a computer for display, and a framegrabber board that can capture and view the video. The SONY PTZ camera attachment costs $3495.

The Sony PTZ Pan-Tilt-Zoom camera attachment for the Pioneer.

PTZ Fast-Track Vision System

This option gives the Pioneer machine vision. It consists of a Newton Labs Cognachrome high-speed visual recognition system, a computer board that can "see" objects by color and shape. It works with the Saphira software to integrate vision into your Pioneer robot (up to three colors), and includes training software. The PTZ Fast-Track Vision System costs $4495.

Deluxe PTZ Video-Vision System

The deluxe version adds real-time video input to the Cognachrome board and to a server for display and exterior processing. It supports tracking of up to six colors and shape recognition. It includes a 2.4 gHz AV Pro

A closeup of the Pioneer vision system board.

An external view of a Pioneer robot with the add-on vision system and visual targets (soda cans).

The view from the robot's onboard camera – what the robot sees. This view is transmitted to a remote display.

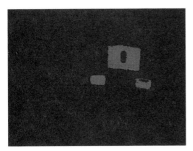

The view from the robot's onboard camera after color segmentation by visual processing.

The scene from the three previous pictures after regional identification. Now the Pioneer can distinguish between the different-colored cans, and knows where they are.

transmitter–receiver set to send the video to the server for display and processing. The Deluxe PTZ Video-Vision System costs $5790.

The following group of pictures from the University of Birmingham's School of Computer Science (Birmingham, UK) show the Fast Track Vision system in action.

See http://www.cs.bham.ac.uk/research/robotics/Pioneer.html for more information. See the Pioneer Web site (http://www.activmedia.com/robots/) for more information on options for their robots.

Other Options

Other options for the Pioneer series include a rear sonar ring, a laser rangefinder, a GPS global positioning system, and a security microphone.

Programming the Pioneer

The Pioneer works with a variety of control software. The most common interfaces are Saphira and P-LOGO.

Saphira

Saphira is an innovative robot control/simulator system. In general, it uses C++ to compile programs for the Pioneer robots (Visual Basic C++ under Windows; other compilers for Unix and variants). You write the programs in C++ and compile them into the Saphira client application, then run that application to connect to the robot. You also use Saphira to modify the robot's behaviors, according to parameters that you define and load into it.

The Pioneer Server application acts as a real-time representation of the robot. It shows a graphical picture of the robot (a small circle), with a direction line to indicate the front. This application, when connected to a Pioneer robot, passes position information and sensor readings to the Saphira client.

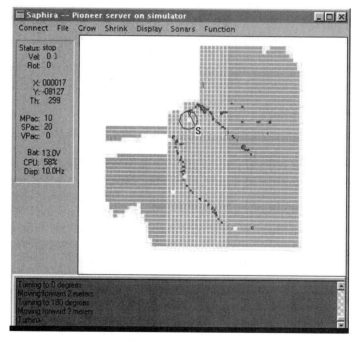

The Saphira application.

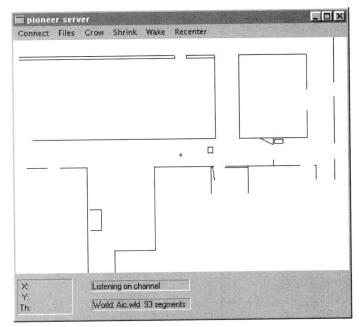

The Pioneer server application.

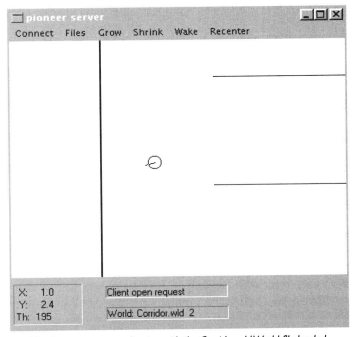

The Pioneer server application with the Corridor.wld World file loaded.

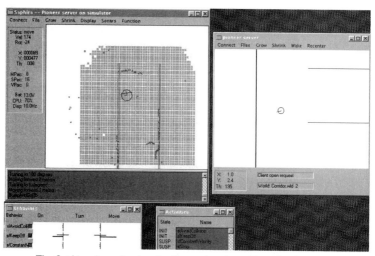

The Saphira client showing the Pioneer locating walls with sonar pings.

The simulator recreates the Pioneer in a software environment, and reacts with it as the Pioneer would. That means that if you load a Corridor world map (see below) into the Server simulator, and connect the Saphira client to it using the Local Port (simulation) option, the robot will react to the corridor walls as the real robot would. That means you can see the robot giving back sonar information on the Server screen.

For the actual robot, you'd connect using TCP (transport control protocol) networking or a serial port cable/radio modem. The Server software will show the real-time position of the Pioneer.

Put them all together, and you get an interactive programming and display environment for your robot. In this example, the Pioneer has

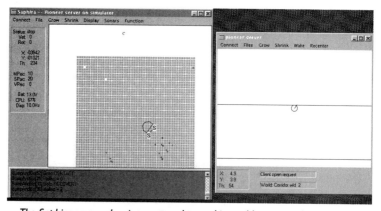

The Saphira screen showing scattered sonar hits and bump switch registration.

been loaded with the Corridor.wld world file, which tells it that it has to navigate a corridor.

The compiled Saphira client is loaded and connected to the robot. In this example, the direct.exe file, found in the /BIN directory, is used. It tells the robot to perform a simple patrolling motion. Note that you can see the robot in the center of the display as it moves and senses the environment. The small dots around the robot are sonar pings. These reflect sound waves off surfaces and back to the robot. The robot collects the information and analyzes it to see where the walls of the corridor are located. The Saphira software, working in conjunction with the Pioneer robot (or the simulator) is capable of making even more sophisticated analysis — locating doors and doorways, for example. See the documentation for more information.

Since one of the behaviors (for this program) loaded into the robot is to avoid walls, the Pioneer will stay between them as it moves about. You can also change the behavior for the robot from the Saphira client, to reprogram it on the fly.

For this example, we canceled the Avoid Walls behavior in the Function/Behaviors dialog box (by clicking on it). Now the Pioneer will run between the corridor walls and use its bump sensors (located around its perimeter) to perform a back-up-and-turn motion (defined as Bump-and-Go in the example program). In the screen shot, note how the Saphira software shows the sonar hits as well as two small "S" indicators that register the wall contact.

A recent addition to the Saphira software is the Colbert robot action evaluator. This is a more straightforward way of writing robot control programs for your Pioneer and running them through Saphira, since you don't have to compile them first. That means you don't need C++ software like Visual Basic C++ just to program the robot. Colbert's structure is similar to Interactive C, and the files are easy to edit in a word processor. Here's how it works:

Open one of the .act files in the Colbert directory of the Saphira distribution. We started with the Sample.act file, which is shown in the illustration at the top of the next page.

Edit the parameters under the relevant sections. For example, you can change the range under the Patrol section, and make the Square into a polygonal motion. Save the file under a different name, and you're ready to put it into Saphira.

To load your new Colbert actions, edit the Init.act file (this should also be in the Colbert directory). Replace the line demo.act with the name of your new .act file. This will now be loaded when you start Saphira, and you should be able to see the changes you made demonstrated.

The Colbert file Sample.act in WordPad.

Changing the Pioneer's Environment

You can also create your own world files to use in your Pioneer. These are text files you can edit in a word processor. The basic format is a set of coordinates. You add walls and doorways to your world by changing the coordinates. You'll have to experiment to see how to match your robot's physical environment in the Saphira software.

This is just a small example of what you can do with the Saphira software. Use the example programs to configure your robot for a variety of sample behaviors. See the documentation for more information. There's a lot of depth to this software, and you can learn a lot from it. See http://www.ai.sri.com/~konolige/saphira/ for more detailed information and software.

P-LOGO

P-LOGO is a variant of the distinguished LOGO programming language, designed to work with the Pioneer. P-LOGO uses simple statements to move a robot "turtle" about on a screen, and to make it perform specific functions. P-LOGO also controls the physical Pioneer robot.

For the Pioneer, P-LOGO replaces Saphira as the client control software (if you so choose). The Pioneer server is run, using the actual robot or the simulator, to connect to the P-LOGO client.

In this figure, the P-LOGO program is making the turtle move in a square. Note that the standard P-LOGO display is used on the client. The Pioneer robot makes the square movement, as shown in the Server window and by the actual performance of the robot.

This is a good way to bridge grade schools and universities. P-LOGO's ease of use makes it a more natural way to control the robot. The P-LOGO programming language also has specific Pioneer commands, so you can experiment with the robot's full range of functions. See the P-LOGO documentation for more information.

A note about Saphira and P-LOGO: Unlicensed software that has both the client and the server-simulator software is available at www.activ media.com/robots. You can download the programs and try the Pioneer programming interfaces on a PC running Windows or Linux, or a Unix workstation. Running the simulator itself is a cool lesson in robotics, though it doesn't match up to the reality of owning a Pioneer.

Alternate higher-level programming interfaces for the Pioneer are also available. Check the web site for more information.

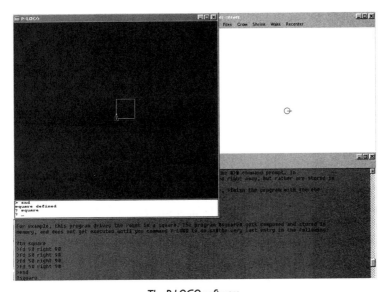

The P-LOGO software.

Further Ideas

There's so much to the Pioneer that it's hard to sum up the ideas of what you can do with it. It has a lot of potential, based on the amount of time you spend learning to program it and the options that you can afford. The Pioneer series can also be programmed using a development environment called Ayllu. This allows for more flexible multi-robot control in different environments. Read more about it at http://robots.activmedia.com/allyu.

Some general suggestions: Use the examples from the other sections of the books for ideas like maze-solving (easy to do with the sonar sensors), and see what you can do with the arbitration between sensors. The wall, doorway, and corridor recognition features are some of the more interesting features of the Saphira software; see how you can relate them to your robot's programming.

You can also look into task-oriented behaviors like delivery under certain conditions. For example, if the Pioneer goes to the end of a corridor and receives a lot of sonar pings from directly in front of it, it can assume the door is closed; if it doesn't, it can assume the door is open, and make the delivery. This kind of program would also use subprogramming like the bump-and-go behavior to avoid getting stuck, the sonar wall-avoidance behavior to navigate, and your own world file as a guide. Finally, add a basket to the flat top of the Pioneer to hold cargo.

Check the documentation for more ideas. It's on the ActivMedia Robotics web site, as noted, and is also listed below.

Final Thoughts

The Pioneer robots are a great attempt to bring professional-grade robotics to a wider market. They are the best-priced robots in their class, with a strong software interface that's developing nicely (Saphira), and a link to the education market (P-LOGO).

The price tag, while acceptable for what you get, may still put the Pioneers out of the running for many school departments and most home hobby experimenters. It would be interesting to see if the price could be brought down, through alternative manufacturing techniques and "costing-down" approaches. This is more a question of industrial design than simply making a "cheaper" robot. Bottom line: This is at least a noble attempt to bring a full-fledged professional robot closer to the education/consumer market.

Contact Information

Pioneer 2-CE
$2295

Pioneer 2-DX
$3295

Pioneer 2-AT
$5495

ActivMedia Robotics
44-46 Concord Street
Peterborough, NH 03458
(603) 924-9100
www.activrobots.com (main Pioneer Web site)
www.mobilerobots.com (Pioneers for Education)

Index

About
the Author

Richard Raucci lives in a beautiful house in San Francisco's Noe Valley with his wife, writer Elizabeth Crane, and their sons, Philip and Michael. A former computer magazine editor, he turned to full-time book writing when the opportunity arose. He has written three books for Springer Verlag on World Wide Web browsers, as well as a guidebook on HTML design for Ziff-Davis Press, and co-authored two Web guides with Yahoo! and IDG Books.

His freelance work has appeared in numerous magazines, including *Publish*, *PCWorld/MultimediaWorld*, *Mac Computing*, *UnixWorld*, *Open Computing*, *Advanced Systems*, *InfoWorld*, *IEEE Computer*, and *The Seybold Report on Internet Publishing*. He also wrote a science-fiction screenplay, Gary Million, which was a Chesterfield Writers Project semi-finalist in 1998.

He is a graduate of the University of Pennsylvania, where he studied dramatic writing with Romulus Linney and earned a degree in English. Past occupations have included Psychiatric Technician (at the Bethesda Naval Hospital), Department of Justice Court Clerk (processing political asylum applications for the Executive Office for Immigration Review in San Francisco), and magazine editor (for IDG and McGraw-Hill). His interests include reading, architecture, photography, design, movies, robotics, technology, science fiction, and the Internet, not necessarily in that order. Reach him at rraucci@well.com.